D0051185

Marine Biology: A Very Short Introduction

VERY SHORT INTRODUCTIONS are for anyone wanting a stimulating and accessible way in to a new subject. They are written by experts, and have been published in more than 25 languages worldwide.

The series began in 1995, and now represents a wide variety of topics in history, philosophy, religion, science, and the humanities. The VSI library now contains more than 300 volumes—a Very Short Introduction to everything from ancient Egypt and Indian philosophy to conceptual art and cosmology—and will continue to grow in a variety of disciplines.

Very Short Introductions available now:

ADVERTISING Winston Fletcher
AFRICAN HISTORY John Parker and
    Richard Rathbone
AGNOSTICISM Robin Le Poidevin
AMERICAN HISTORY Paul S. Boyer
AMERICAN IMMIGRATION
    David A. Gerber
AMERICAN POLITICAL PARTIES
    AND ELECTIONS L. Sandy Maisel
AMERICAN POLITICS
    Richard M. Valelly
THE AMERICAN PRESIDENCY
    Charles O. Jones
ANAESTHESIA Aidan O'Donnell
ANARCHISM Colin Ward
ANCIENT EGYPT Ian Shaw
ANCIENT GREECE Paul Cartledge
ANCIENT PHILOSOPHY Julia Annas
ANCIENT WARFARE
    Harry Sidebottom
ANGELS David Albert Jones
ANGLICANISM Mark Chapman
THE ANGLO-SAXON AGE John Blair
THE ANIMAL KINGDOM
    Peter Holland
ANIMAL RIGHTS David DeGrazia
THE ANTARCTIC Klaus Dodds
ANTISEMITISM Steven Beller
ANXIETY Daniel Freeman and
    Jason Freeman
THE APOCRYPHAL GOSPELS
    Paul Foster
ARCHAEOLOGY Paul Bahn
ARCHITECTURE Andrew Ballantyne

ARISTOCRACY William Doyle
ARISTOTLE Jonathan Barnes
ART HISTORY Dana Arnold
ART THEORY Cynthia Freeland
ATHEISM Julian Baggini
AUGUSTINE Henry Chadwick
AUSTRALIA Kenneth Morgan
AUTISM Uta Frith
THE AVANT GARDE
    David Cottington
THE AZTECS David Carrasco
BACTERIA Sebastian G. B. Amyes
BARTHES Jonathan Culler
BEAUTY Roger Scruton
BESTSELLERS John Sutherland
THE BIBLE John Riches
BIBLICAL ARCHAEOLOGY
    Eric H. Cline
BIOGRAPHY Hermione Lee
THE BLUES Elijah Wald
THE BOOK OF MORMON
    Terryl Givens
BORDERS Alexander C. Diener and
    Joshua Hagen
THE BRAIN Michael O'Shea
THE BRITISH CONSTITUTION
    Martin Loughlin
BRITISH POLITICS Anthony Wright
BUDDHA Michael Carrithers
BUDDHISM Damien Keown
BUDDHIST ETHICS Damien Keown
CANCER Nicholas James
CAPITALISM James Fulcher
CATHOLICISM Gerald O'Collins

Available soon:

For more information visit our website
www.oup.com/vsi/

Philip V. Mladenov

MARINE
BIOLOGY

A Very Short Introduction

OXFORD
UNIVERSITY PRESS

OXFORD

UNIVERSITY PRESS

Great Clarendon Street, Oxford, OX2 6DP,
United Kingdom

Oxford University Press is a department of the University of Oxford.
It furthers the University's objective of excellence in research, scholarship,
and education by publishing worldwide. Oxford is a registered trade mark of
Oxford University Press in the UK and in certain other countries

Published in the United States of America by Oxford University Press
198 Madison Avenue, New york, NY 10016, United States of America

British Library Cataloguing in Publication Data

Data available

ISBN 978-0-19-969505-8

Printed in Great Britain by
Ashford Colour Press Ltd, Gosport, Hampshire

*For my family, both in Canada
and New Zealand*

Contents

Acknowledgements

I am grateful to Latha Menon and Emma Ma for their advice and encouragement throughout this project and to Paul Tyler for his careful reading and astute comments on the manuscript.

List of illustrations

Marine Biology

List of tables

List of abbreviations

CFCs	chlorofluorocarbons
DOC	dissolved organic carbon
DOM	dissolved organic matter
DSL	deep scattering layer
EEZ	exclusive economic zone
ENSO	El Niño Southern Oscillation
HAB	harmful algal bloom
HNLC	high nutrient-low chlorophyll
POM	particulate organic matter
ppm	parts per million
UV-B	ultraviolet-B

Introduction

The marine environment is the planet's largest and most important habitat. It covers 71 per cent of the planet's surface; contains more than 99 per cent of its habitable living space; represents its largest repository of living matter; produces half its primary production; and supports a remarkably diverse and exquisitely adapted array of life forms, from microscopic viruses, bacteria, and plankton to the largest existing animals. For these reasons alone, marine biology is an inherently fascinating, important, and exciting subject.

However, a solid understanding of marine biology is now more relevant and timely than ever before. This is because humans are beginning to fully appreciate that the marine environment and the organisms that live in that environment are critical to our well-being and survival. The marine system of our planet provides us and our livestock with a rich source of food; it helps stabilize our climate, which is of great importance as human-induced climate change accelerates during this century; it absorbs many of our waste products; it provides us with a wide range of biomolecules of importance in medicine and engineering; properly functioning coral reef and mangrove systems protect our coastlines; and marine ecosystems support recreation and tourism throughout the world.

Unfortunately, human activity has had a severe impact on this vast habitat for many years. As the 21st century progresses, it has become clear that human activities such as overfishing, coastal development, sewage disposal, plastic pollution, oil spills, nutrient pollution, the spread of exotic species, and the emission of climate-changing greenhouse gases are causing significant changes and damage to the marine environment and to many of the life forms living in that environment. As global population rises exponentially from seven to nine billion over the next forty years, such pressures will inevitably intensify and will pose a serious threat to human welfare.

This book aims to provide a brief but holistic introduction to the marine environment and the nature of life in the oceans so that readers can fully appreciate the inherent beauty and complexity of the marine environment, its significance to the planet and to human society, and some of the consequences of increasing human impacts on the oceans.

Chapter 1
The marine environment

Viewed from space, our planet is clearly dominated by its greatest natural feature—a vast, deep, and interconnected mass of seawater—the Global Ocean. The Global Ocean contains an enormous amount of water—about 1.34 billion cubic kilometres—which constitutes about 97 per cent of all the water that exists on the planet.

Geography of the Global Ocean

The Global Ocean has come to be divided into five regional oceans—the Pacific, Atlantic, Indian, Arctic, and Southern Oceans (see Figure 1). Some of the marginal areas of these regional oceans are familiarly known to us as seas, for example the Caribbean Sea or the Red Sea.

These oceans are large, seawater-filled basins that share characteristic structural features (see Figure 2). The edge of each basin consists of a shallow, gently sloping extension of the adjacent continental land mass and is termed the continental shelf or continental margin. Continental shelves typically extend offshore to depths of a couple of hundred metres and vary from several kilometres to hundreds of kilometres in width.

Marine Biology

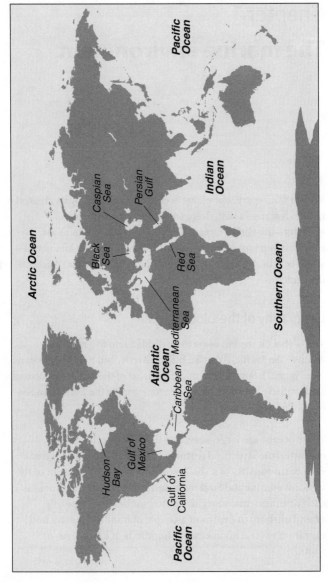

1. The Global Ocean

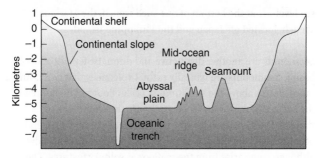

2. Diagrammatic cross-section of an ocean basin

At the outer edge of the continental shelf, the seafloor drops off abruptly and steeply to form the continental slope, which extends down to depths of 2–3 kilometres. The continental slope then flattens out and gives way to a vast expanse of flat, soft, ocean bottom—the abyssal plain—which extends over depths of about 3–5 kilometres and accounts for about 76 per cent of the Global Ocean floor.

The abyssal plains are transected by extensive mid-ocean ridges—underwater mountain chains created by intense volcanic activity—that can rise thousands of metres or more above the surrounding abyssal plains. Mid-ocean ridges form a continuous chain of mountains that extend linearly for 65,000 kilometres across the floor of the Global Ocean basins—akin to the seams on a baseball.

In some places along the edges of the abyssal plains the ocean bottom is cut by narrow, oceanic trenches or canyons which plunge to extraordinary depths—3–4 kilometres below the surrounding seafloor—and are thousands of kilometres long but only tens of kilometres wide. In comparison, the Grand Canyon is about 1.6 kilometres deep and 433 kilometres long. The deepest known part of the Global Ocean—at around 11 kilometres below sea level—is at the bottom of one such trench, the Mariana Trench, located off Japan and the Philippine Islands.

Seamounts are another distinctive and dramatic feature of ocean basins. Seamounts are typically extinct volcanoes that rise 1,000 or more metres above the surrounding ocean floor but do not reach the surface of the ocean. Their peaks are thus hundreds to thousands of metres below the sea surface. Seamounts generally occur in chains or clusters in association with mid-ocean ridges, although some arise from the seafloor as solitary features. The Global Ocean contains an estimated 100,000 or so seamounts that rise more than 1,000 metres above the surrounding deep-ocean floor.

Environmental conditions in the oceans

Marine organisms live throughout the Global Ocean from the surface to the bottom of its deepest trenches, suspended or actively swimming in open-ocean water, termed the pelagic zone, or living on or within the ocean bottom, the benthic zone.

This colonization of every part of the marine environment by living things is now taken for granted, although in the 19th century it was widely believed that no life existed in what was then termed the 'azoic zone'—any part of the oceans beneath 300 fathoms (about 550 metres). Here the environment was considered a dead zone of darkness, too inhospitable for any life form. This notion was firmly laid to rest following the historic expedition of HMS *Challenger* (1872–76), the first ship to comprehensively explore the deeper parts of ocean basins and to discover marine life to depths of close to 6,000 metres.

We now know that the oceans are literally teeming with life. Viruses, the most primitive of life forms, are astoundingly abundant—there are around ten million viruses per millilitre of seawater. Bacteria and other microorganisms occur at concentrations of around 1 million per millilitre; and there are hundreds of thousands of different species of invertebrates, fish, marine mammals, and marine reptiles living in all parts of the Global Ocean.

So what sort of environment does this plethora of life occupy?

Salinity

The water in the oceans is in the form of seawater, a dilute brew of dissolved ions, or salts, which bathes all living marine organisms. Chloride and sodium ions are the predominant salts in seawater, along with smaller amounts of other ions such as sulphate, magnesium, calcium, and potassium (see Table 1).

The total amount of dissolved salts in seawater is termed its salinity. Seawater typically has a salinity of roughly 35—equivalent to about 35 grams of salts in one kilogram of seawater. However, this can vary, particularly in partially enclosed bays subject to high rates of evaporation, precipitation, river run off, or ice melt.

Table 1 Major ions in seawater

Ion	Weight (grams/kilograms)
Chloride (Cl^-)	19.35
Sodium (Na^+)	10.76
Sulphate (SO_4^{2-})	2.71
Magnesium (Mg^{2+})	1.29
Calcium (Ca^{2+})	0.41
Potassium (K^+)	0.39
Total	**34.91**

Temperature

Most marine organisms are exposed to seawater that, compared to the temperature extremes characteristic of terrestrial environments, ranges within a reasonably moderate range. Surface waters in tropical parts of ocean basins are consistently warm throughout the year, ranging from about 20–27°C, and up to around 30°C in shallow tropical bays at the height of summer. On the other hand, surface seawater in polar parts of ocean basins can get as cold as –1.9°C.

Sea temperatures typically decrease with depth, but not in a uniform fashion. A distinct zone of rapid temperature transition is often present that separates warm seawater at the surface from cooler deeper seawater. This zone is called the thermocline layer (see Figure 3).

In tropical ocean waters the thermocline layer is a strong, well-defined and permanent feature. It may start at around 100 metres and be a hundred or so metres thick. Sea temperatures above the thermocline can be a tropical 25°C or more, but only 6–7°C just below the thermocline. From there the temperature drops very gradually with increasing depth. Thermoclines in temperate ocean regions are a more seasonal phenomenon, becoming well established in the summer as the sun heats up the surface waters, and then breaking down in the autumn and winter. Thermoclines are generally absent in the polar regions of the Global Ocean.

Human-induced global climate change, which is resulting in increasing average global air temperatures, is also resulting in rising ocean temperatures. The average temperature of the Global Ocean is about 1°C higher now than 140 years ago and will continue to rise throughout this century at least. It is projected that warming will be greatest in the upper 100 metres of the oceans in the early part of this century. After this,

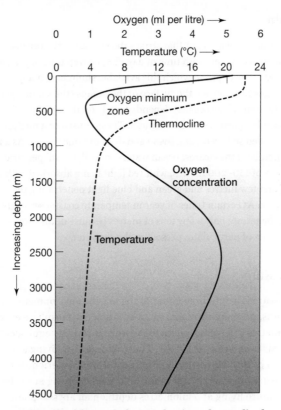

Oxygen (ml per litre) →

| 0 | 1 | 2 | 3 | 4 | 5 | 6 |

Temperature (°C) →

| 0 | 4 | 8 | 12 | 16 | 20 | 24 |

Oxygen minimum zone

Thermocline

Oxygen concentration

Temperature

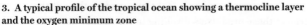

3. **A typical profile of the tropical ocean showing a thermocline layer and the oxygen minimum zone**

temperatures will begin to increase in deeper parts of the oceans as warmer surface seawater is slowly mixed to deeper depths. This warming trend is causing a rapid decrease in the thickness and coverage of sea ice in the Arctic Ocean, leaving increasingly larger areas of ice-free open water each summer. It is also impacting on marine organisms and the functioning of marine ecosystems in a variety of important ways that we will explore throughout this volume.

Light

The amount of sunlight that strikes the surface of the oceans varies considerably with time of day, cloud cover, time of year, and latitude. The depth to which this available sunlight actually manages to penetrate the surface of the oceans (the sunlit layer or photic zone of the oceans) depends largely on the amount of suspended particles in the seawater. These consist of a mixture of suspended sediment, and living and dead organic matter. As a rule of thumb, in the clearest ocean waters some light will penetrate to depths of 150–200 metres, with red light being absorbed within the first few metres and green and blue light penetrating the deepest. At certain times of year in temperate coastal seas light may penetrate only a few tens of metres because of the large amounts of particulates present in the seawater.

Pressure

Pressure is a defining feature of the marine environment. In the oceans, pressure increases by an additional atmosphere every 10 metres (one atmosphere of pressure is roughly equivalent to the air pressure at sea level). Thus, an organism living at a depth of 100 metres on the continental shelf experiences a pressure ten times greater than an organism living at sea level; a creature living at 5 kilometres depth on an abyssal plain experiences pressures some 500 times greater than at the surface; those organisms dwelling in the deeper parts of oceanic trenches are subject to pressures some 1,000 times greater than a sea-level dweller—the pressure at these depths equates to a massive 10,000 tonnes per square metre.

Oxygen

With very few exceptions, dissolved oxygen is reasonably abundant throughout all parts of the Global Ocean. However, the amount of oxygen in seawater is much less than in air—seawater at 20°C contains about 5.4 millilitres of oxygen per litre of seawater, whereas air at this temperature contains about 210 millilitres of oxygen per litre. The colder the seawater, the more oxygen it contains; for example, seawater at 0°C contains around 8 millilitres of oxygen per litre.

Oxygen is not distributed evenly with depth in the oceans. Oxygen levels are typically high in a thin surface layer 10–20 metres deep. Here oxygen from the atmosphere can freely diffuse into the seawater, plus there is plenty of floating plant life producing oxygen through photosynthesis. Oxygen concentration then decreases rapidly with depth and reaches very low levels, sometimes close to zero, at depths of around 200–1,000 metres. This region is referred to as the oxygen minimum zone (see Figure 3). This zone is created by the low rates of replenishment of oxygen diffusing down from the surface layer of the ocean, combined with the high rates of depletion of oxygen by decaying particulate organic matter that sinks from the surface and accumulates at these depths.

Beneath the oxygen minimum zone, oxygen content increases again with depth such that the deep oceans contain quite high levels of oxygen, though not generally as high as in the surface layer. The higher levels of oxygen in the deep oceans reflect in part the origin of deep-ocean seawater masses, which are derived from cold, oxygen-rich seawater that sinks rapidly down from the surface of polar oceans, thereby conserving its oxygen content. As well, compared to life in near-surface waters, organisms in the deep ocean are comparatively scarce and have low metabolic rates, thus consuming little of the available oxygen.

Carbon dioxide and ocean acidification

In contrast to oxygen, carbon dioxide (CO_2) dissolves readily in seawater. Some of it is then converted into carbonic acid (H_2CO_3), bicarbonate ion (HCO_3^-), and carbonate ion (CO_3^{2-}), with all four compounds existing in equilibrium with one another, as shown in the following equilibrium equation:

$$CO_2 + H_2O \leftrightarrow H_2CO_3 \leftrightarrow H^+ + HCO_3^- \leftrightarrow H^+ + CO_3^{2-}$$

The pH of seawater is inversely proportional to the amount of carbon dioxide dissolved in it. Referring back to the equilibrium equation, it can be seen that the more carbon dioxide absorbed by the oceans, the more the equilibrium shifts to the right, releasing more hydrogen ions (H^+), which lowers the pH.

Seawater is naturally slightly alkaline, with a pH ranging from about 7.5 to 8.5, and marine organisms have become well adapted to life within this stable pH range. Seawater near the surface of oceans is generally at the higher end of this range because marine plant life is taking up carbon dioxide for photosynthesis, thereby shifting the equilibrium reaction to the left and removing hydrogen ions. Furthermore, surface seawater is generally warmer than deep-ocean water and the warmer the seawater, the less carbon dioxide it can absorb. In deep, colder parts of the oceans, where no photosynthesis is taking place, carbon dioxide concentrations are higher and the pH is at the lower end of this range.

As a result of this carbonic acid–bicarbonate–carbonate equilibrium, the Global Ocean is a vast reservoir of planetary carbon—which has important implications for marine life and human society. In the oceans, carbon is never a limiting factor to marine plant photosynthesis and growth, as it is for terrestrial plants. From a planetary perspective, the Global Ocean is an enormous natural sink for atmospheric carbon dioxide, a climate-changing greenhouse gas. At present, the Global Ocean is absorbing about 25 per cent of the roughly 35 billion tonnes of carbon dioxide being spewed into the atmosphere each year by humans burning fossil fuels and deforesting the planet's surface—this is roughly one million tonnes of carbon dioxide per hour. Another 25 per cent or so is absorbed by forests, with the balance accumulating in the atmosphere. The net result is that atmospheric carbon dioxide concentrations are currently rising at a rate of about 2 parts per million (ppm) per year, which is why carbon dioxide concentrations in the planet's atmosphere have increased from a pre-industrial level of 278 ppm to a level now greater than 380 ppm and rising.

Thus the oceans play a fundamental role in damping the rate of human-induced climate change, but they are starting to show the effects of their mopping up excess carbon dioxide from the atmosphere for the last two and a half centuries. The pH of the oceans is now becoming more acidic. Since the beginning of the industrial revolution, the average pH of the Global Ocean has dropped by about 0.1 pH unit, making it 30 per cent more acidic than in pre-industrial times. This is a large change over such a short period of time, and this will accelerate at present rates of carbon dioxide emissions—by 2065 the average pH of the oceans is forecast to drop by another 0.15 units.

As a result, more and more parts of the oceans are falling below a pH of 7.5 for longer periods of time. This trend, termed ocean acidification, is having profound impacts on marine organisms and the overall functioning of the marine ecosystem. For example, many types of marine organisms such as corals, clams, oysters, sea urchins, and starfish manufacture external shells or internal skeletons containing calcium carbonate. When the pH of seawater drops below about 7.5, calcium carbonate starts to dissolve, and thus the shells and skeletons of these organisms begin to erode and weaken, with obvious impacts on the health of the animal. Also, these organisms produce their calcium carbonate structures by combining calcium dissolved in seawater with carbonate ion. As the pH decreases, more of the carbonate ions in seawater become bound up with the increasing numbers of hydrogen ions, making fewer carbonate ions available to the organisms for shell-forming purposes. It thus becomes more difficult for these organisms to secrete their calcium carbonate structures and grow.

The oceans in motion

On a planetary scale, the surface of the Global Ocean is moving in a series of enormous, roughly circular, wind-driven current systems, or gyres, each thousands of kilometres in diameter (see Figure 4). The northern hemisphere gyres in the North Pacific and North Atlantic Oceans flow clockwise; the southern hemisphere gyres in the South Pacific, South Atlantic, and the Indian Oceans flow counterclockwise. These gyres transport enormous volumes of water and heat energy from one part of an ocean basis to another, and carry along many kinds of non-swimming, or poor-swimming, forms of marine life, called plankton.

The North Atlantic Ocean gyre system provides a good example of the dynamics of gyre systems. Here the surface waters circulate in typical clockwise fashion. The northward-flowing western edge of this gyre comprises the Gulf Stream. The Gulf Stream is a 50–75-kilometre-wide fast-moving surface current that transports vast volumes of warm, salty tropical seawater at speeds averaging 3–4 kilometres per hour up along the eastern edge of the North American continent. This warm current leaves the coast of North America at around South Carolina and traverses the North Atlantic as the North Atlantic Drift Current, releasing its heat into the atmosphere off the coast of northern Europe and moderating the European climate. The gyre then turns southward and meanders down along the western edge of Europe and Africa as the cooler, broader, and more slowly moving Canary Current. This current then curves westward along the equator to form the North Equatorial Current which flows into the Caribbean region to complete the gyre.

In the centre of each oceanic gyre is a large pool of stable, clear seawater. In the North Atlantic Ocean this region is referred to as the Sargasso Sea, which is roughly 3,000 kilometres long and 1,000 kilometres wide.

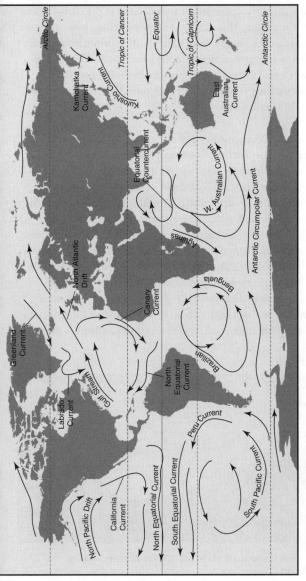

4. **The major surface currents of the Global Ocean**

Beneath the surface, the deeper water masses of the Global Ocean are also in motion. This motion is not wind driven, however, as for surface currents, but is a more stately motion driven by buoyancy changes in seawater occurring in the polar oceans. This kind of flow is referred to as thermohaline circulation because the buoyancy changes are a result of changes in the temperature and salinity of the seawater.

The Atlantic Ocean provides a good example of thermohaline circulation. The Gulf Stream transports large amounts of warm, salty seawater from the tropics to polar latitudes in the North Atlantic Ocean. Here it is cooled and its salinity increased by the addition of salt extruded from the freezing of Arctic seawater. This process creates cold, salty, and very dense seawater that sinks rapidly to great depths. Large amounts of such seawater are formed and sink in the Norwegian Sea off Iceland and Greenland, forming what is called the North Atlantic Deep Water (see Figure 5). From there it flows slowly southward near the bottom of the Atlantic Ocean basin to emerge hundreds of years later near the coast of Antarctica. Similarly, cold, salty, and dense seawater also sinks off the coast of Antarctica and flows northwards just beneath the

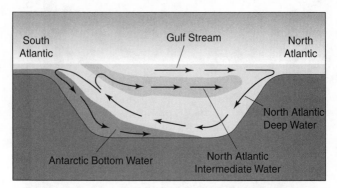

5. Simplified schematic of the Atlantic Ocean thermohaline circulation

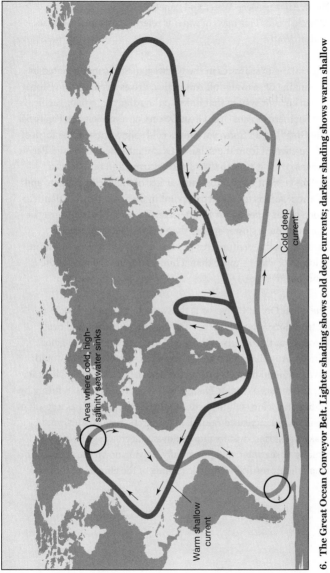

6. **The Great Ocean Conveyor Belt. Lighter shading shows cold deep currents; darker shading shows warm shallow currents. The circles show areas of sinking cold, high-salinity seawater**

Area where cold, high-salinity seawater sinks

Cold deep current

Warm shallow current

North Atlantic Deep Water and penetrates far into the North Atlantic basin. This mass of water is referred to as the Antarctic Bottom Water.

The sinking of seawater in the Norwegian Sea, supplemented by the sinking of seawater off Antarctica, drives a remarkable Global Ocean current system that links up the Atlantic, Indian, Pacific, and Southern Oceans—the Great Ocean Conveyor Belt (see Figure 6). The Great Ocean Conveyor moves cold saline water in the form of a deep current from the Atlantic Ocean into the Indian and Pacific Oceans. Here it rises to the surface, warms, and returns as a surface current back across the surface of the Pacific, Indian, and Atlantic Oceans to its starting point in the polar North Atlantic, a process that takes about 1,000 years. The Great Ocean Conveyor moves at much slower speeds than the wind-driven surface currents—a few centimetres per second—but it moves enormous volumes of water—more than a hundred times the flow of the Amazon River.

The Great Ocean Conveyor is thus akin to a planetary-scale distribution system, transporting oxygen, nutrients, and heat throughout the oceans of the world and moderating the global climate. Any change in the characteristics of this system would have profound impacts on climate and society. It is possible that human-induced climate change will affect this system in future by adding more low-salinity water to the Norwegian Sea as a result of ice melt and increased rainfall in the Arctic region. This would make the surface waters more buoyant, decreasing the rate of sinking of seawater, and potentially slowing down the system. This would likely result in a drastic cooling of the climate of Europe.

Chapter 2
Marine biological processes

Roughly half of the planet's primary production—the synthesis of organic compounds by chlorophyll-bearing organisms using energy from the sun—is produced within the Global Ocean. On land the primary producers are large, obvious, and comparatively long-lived—the trees, shrubs, and grasses characteristic of the terrestrial landscape. The situation is quite different in the oceans where, for the most part, the primary producers are minute, short-lived microorganisms suspended in the sunlit surface layer of the oceans. These energy-fixing microorganisms—the oceans' invisible forest—are responsible for almost all of the primary production in the oceans. This energy is then transferred to and sustains all of the other organisms in the marine system. These tiny photosynthetic creatures are thus the fundamental drivers of the economy of the Global Ocean.

Marine primary producers

A large amount, perhaps 30–50 per cent, of marine primary production is produced by bacterioplankton comprising tiny marine photosynthetic bacteria ranging from about 0.5 to 2 µm in size. These bacteria are found everywhere in the oceans in great abundance—some are free-living, planktonic forms; others are attached to small particles suspended in the seawater. Much is yet to be learned about this important group of marine organisms, but it appears that one group of photosynthetic bacteria, the cyanobacteria, or blue-green bacteria, may be the most abundant and productive photosynthetic organisms on the planet, occurring in the top 100 metres or so of the oceans at concentrations of around one million cells per millilitre.

Another major group of marine primary producers is the phytoplankton, a diverse group of single-celled organisms about 2 to 200 μm in size. Diatoms are an important group of phytoplankters. Each diatom cell is enclosed within an ornately sculptured, clear glass box made of silica, called a frustule (see Figure 7a). The individual cells of some species of diatom can link up to form colonies of long chains. In some places on the bottom of the oceans, the sediments are composed largely of diatom frustules that have sunk to the seabed in large numbers over long periods of time to create diatomite sediments. Diatomite deposits found on land are derived from uplifted diatomite sediments and are of economic importance, being used in filters, absorbents, as mild abrasives in products such as toothpaste, and in numerous other applications.

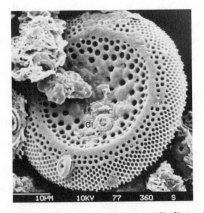

7a. Scanning electron micrograph of a diatom (di=diatom)

7b. Scanning electron micrograph of a dinoflagellate

The dinoflagellates are another major group of single-celled, microscopic phytoplankters. Dinoflagellate cells possess two hair-like flagella, providing them with some limited motility, and are often armoured with translucent plates made of cellulose (see Figure 7b).

Coccolithophores are yet another important group of microscopic, single-celled phytoplankters. Each cell is covered with small ornamented plates, called coccoliths, made of calcium carbonate (see Figure 7c). Coccoliths are responsible for the formation of chalk deposits in deep-ocean sediments. Uplifted chalk deposits on land are mined to produce products such as quicklime and slaked lime which are used in a wide range of industrial processes.

Marine primary production

Bacterioplankton and phytoplankton use the chlorophyll pigments in their cells to harvest the energy of sunlight penetrating the photic zone of the oceans. Through the process of photosynthesis this energy is used to synthesize energy-rich

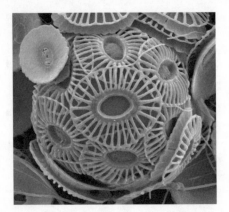

7c. Scanning electron micrograph of a coccolithophore

carbon containing organic compounds, such as sugars and amino acids. Dissolved carbon dioxide (CO_2) in the seawater provides the source of inorganic carbon for this process. In the presence of light energy, carbon dioxide reacts with water to produce organic compounds, such as glucose. Oxygen (O_2) is a by-product of this process and is released into the surrounding seawater.

The general equation for photosynthesis is:

$$6CO_2 + 6H_2O \xrightarrow{\text{Sunlight}} \underset{\substack{\text{organic} \\ \text{compounds}}}{C_6H_{12}O_6} + 6O_2$$

The organic matter produced by these primary producers is the energy base, or first trophic level, of the Global Ocean; it represents the major source of energy supporting life in the oceans and it drives the marine ecosystem as whole. The energy at the first trophic level is utilized by a diversity of marine organisms at the second trophic level—the primary consumers or herbivores—which are in turn eaten by marine consumers at higher trophic levels in the system.

The rate of photosynthesis decreases with depth in the oceans because of decreasing light intensity. Since the upper layers of the oceans are a naturally turbulent environment, phytoplankton is mixed to various depths within the water column depending on the strength of vertical circulation. If a phytoplankton cell is to grow and reproduce it must spend enough time above a certain depth, often referred to as the 'critical depth', to be able to photosynthesize more energy than is required for its basic metabolic requirements. Otherwise all of the energy produced is respired and there is nothing left for growth. Thus, light availability and the strength of vertical mixing are important factors limiting primary production in the oceans.

Nutrient availability is the other main factor limiting the growth of primary producers. One important nutrient is nitrogen, which plants require for a variety of metabolic functions. In particular, nitrogen is a key component of amino acids, which are the building blocks of proteins. Nitrogen is absorbed by marine photosynthetic organisms in the form of dissolved ammonium (NH_4^+), nitrite (NO_2^-) or nitrate (NO_3^-). Nitrogen-'fixing' bacteria provide a natural source of these nitrogen-based inorganic compounds in the oceans. These bacteria are able to produce or 'fix' these compounds from molecular nitrogen (N_2) dissolved in the seawater; they are then released into the seawater when the bacteria die and become available to primary producers.

Photosynthetic marine organisms also need phosphorus, which is a requirement for many important biological functions, including the synthesis of nucleic acids, a key component of DNA. Phosphorus in the oceans comes naturally from the erosion of rocks and soils on land, and is transported into the oceans by rivers, much of it in the form of dissolved phosphate (PO_4^{3-}), which can be readily absorbed by marine photosynthetic organisms.

Once these inorganic nitrogen and phosphorus compounds become incorporated into large organic molecules within marine primary producers, they become available to other organisms at higher trophic levels in the marine system when they consume these organisms.

Inorganic nitrogen and phosphorus compounds are abundant in deep-ocean waters. Here, a constant rain of dead organic material sinking down from the surface waters is decomposed by bacteria, recycling inorganic nitrogen and phosphorus compounds back into the seawater. When the upper layer of the ocean is well mixed, or unstratified, these deep nutrient-rich waters are mixed up into the photic zone, bringing an abundant supply of nutrients to the surface. However, when a thermocline is present, it acts as a barrier to the regeneration of nutrients from the deep-oceanic water below. Under such circumstances, and if light levels are not limiting, photosynthetic organisms will rapidly deplete nutrients from the surface layer above the thermocline. In practice, inorganic nitrogen and phosphorus compounds are not used up at exactly the same rate. Thus one will be depleted before the other and becomes the limiting nutrient at the time, preventing further photosynthesis and growth of marine primary producers until it is replenished.

Nitrogen is often considered to be the rate-limiting nutrient in most oceanic environments, particularly in the open ocean. However, in coastal waters phosphorus is often the rate-limiting nutrient. It also appears that nitrogen and phosphorus can work synergistically in the marine environment so that if both are abundant, then the rate of primary production is higher than if just one or the other is abundant.

Marine primary producers also require the essential micronutrient iron, which helps plants utilize the nitrate that they require for growth. The iron in the oceans is derived from iron-rich dust that is blown far out into the oceans from deserts during dust storms. Iron deposits at the edge of continents are another source. Adequate concentrations of dissolved iron exist in most parts of the Global Ocean, so it is not normally a limiting factor for primary production. However, in a few open-ocean regions, such as in the eastern equatorial Pacific and parts of the Southern Ocean, concentrations of dissolved iron are so low that it becomes the rate-limiting factor for primary production despite high levels of nitrogen and phosphorus being present. Such areas are referred to as high nutrient-low chlorophyll (HNLC) regions, the low chlorophyll concentrations reflecting the dearth of chlorophyll-containing organisms in the seawater.

Some researchers have suggested that artificially fertilizing these HNLC regions with iron to kick-start primary production in these otherwise nutrient-rich regions could be a way of mitigating climate change. The idea is that the resultant bloom in primary producers will draw down large amounts of carbon dioxide from the atmosphere. When these organisms die they will sink to the ocean floor where the carbon in their tissues will be locked away in marine sediments. In other words, the primary producers are 'pumping' carbon dioxide from the atmosphere into long-term storage in deep-ocean sediments.

Indeed, small-scale trials involving fertilizing small patches of the Southern Ocean with several tonnes of dissolved iron from research vessels have shown that primary production can be stimulated in this way. Whether this will ever be a practical way to geo-engineer the large-scale removal of carbon dioxide from the atmosphere is very uncertain. To work well the process will have to stimulate the right kind of phytoplankton blooms—large diatom cells that are resistant to being grazed by organisms at the second trophic level, and that sink rapidly to the sea floor. Otherwise the primary producers are consumed by zooplankton well before they reach the seabed where the carbon can be sequestered. The trials so far have shown that this does not always happen. Also, it is not entirely clear if enough carbon can be transported to the seabed via phytoplankton to actually make a difference. And finally, no one knows what the impacts might be on the greater marine biological system of adding large quantities of iron to the oceans. Thus, although this line of research is furthering our understanding of primary production processes in the oceans, it remains doubtful that it will play a role in limiting human-induced climate change.

Measuring marine primary productivity

The rate of primary production—termed primary productivity—varies considerably over space and time in the Global Ocean. Primary productivity is often expressed as the number of grams of carbon (C) 'fixed', or incorporated into organic matter, per square metre of ocean surface per year (g C m^{-2} yr^{-1}).

Measuring primary productivity in the oceans is a challenging business. It was first done using the light and dark bottle method, which is based on the principle of measuring the amount of oxygen in a sample of seawater after it has been put into a bottle and exposed to light for a period of time, compared to the amount of oxygen in a similar sample of seawater put into an identical bottle under identical conditions, except that the bottle has been covered to exclude all light. Under these conditions, the microscopic primary producers in the light bottle are photosynthesizing and releasing oxygen (with some of this oxygen consumed by respiration), whereas the primary producers in the dark bottle cannot photosynthesize, but do consume some oxygen for respiration. The difference in oxygen content measured between the two bottles at the end of the experiment can be converted into an estimate of the amount of carbon that has been fixed into organic material. This is because the number of molecules of oxygen produced is equivalent to the number of molecules of carbon dioxide fixed into organic matter (see the general equation for photosynthesis, page 22). Using this approach, an estimate of primary productivity in a particular patch of ocean in a particular season can be obtained.

As it turns out, the light and dark bottle method is not sensitive enough to give good results in the large areas of the ocean with low levels of primary productivity. This led to the development of the carbon-14 method for estimating marine primary productivity. In this method, a tiny but known amount of radioactive carbon-14 is added to a bottle containing a sample of seawater, and the bottle is exposed to light for a period of time. In practice, samples of seawater taken from various depths in the photic zone are collected and put into bottles which are exposed to light or 'incubated' on board a research vessel. The light intensity is adjusted with filters to simulate the light intensity at the depth from which the samples were taken.

During this incubation period, some of the carbon-14 in the bottle, along with non-radioactive carbon-12 naturally present in the seawater sample, will be fixed into organic material by the photosynthetic organisms in the bottle. The carbon-14 thus acts as a tracer for the amount of non-radioactive carbon that has been fixed by the photosynthetic organisms in the bottle. At the end of the incubation period, the seawater in the bottle is passed through a fine filter that retains most of the photosynthetic organisms in the sample and the amount of radioactivity on the filter is measured. Since the amount of carbon-14 measured on the filter is proportional to the amount of non-radioactive carbon-12 fixed by the photosynthetic organisms, an estimate of primary productivity can be obtained.

The problem with *in situ* techniques like the light and dark bottle and carbon-14 techniques is that, even with the most intensive sampling regime, they can only provide estimates of primary productivity for a small number of locations at any particular time. It is therefore difficult to build up a global picture of patterns of primary productivity using this approach alone. This changed when satellite observations of the colour of surface seawater became available in the late 1990s. Such satellites are now used routinely to estimate chlorophyll concentrations over very large areas of the Global Ocean—the greener the surface seawater, the greater the chlorophyll concentration, and hence the more photosynthetic organisms present. Repeated satellite-derived measurements of Global Ocean colour over scales of days, weeks, and years, coupled with direct, *in situ* measurements have provided a much improved synoptic understanding of spatial and temporal patterns in global primary production.

Patterns of marine primary productivity

The overall pattern of primary production in the Global Ocean depends greatly on latitude (see Figure 8). In polar oceans primary production is a boom-and-bust affair driven by light availability. Here the oceans are well mixed throughout the year so nutrients are rarely limiting. However, during the polar winter there is no light, and thus no primary production is taking place. In the spring, light levels and day length both increase rapidly and a point is reached during the year in which both nutrients and light become simultaneously non-limiting, and an intense bloom of primary production commences. This may last for several months until light once again becomes limiting in the autumn. Although limited to a short seasonal pulse, the total amount of

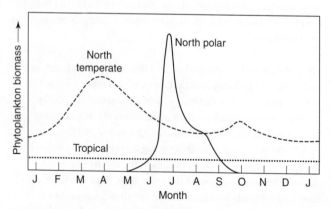

8. **Seasonal variation in phytoplankton biomass in temperate, tropical, and polar seas**

primary production can be quite high, especially in the polar Southern Ocean where annual productivity can be in the order of 100 g C m^{-2} yr^{-1} and in some areas much more.

In tropical open oceans, primary production occurs at a low level throughout the year. Here light is never limiting but the permanent tropical thermocline prevents the mixing of deep, nutrient-rich seawater with the surface waters. Hence nutrients are present at permanently low levels in the photic zone, which limits primary productivity to low levels throughout the year. Hence, open-ocean tropical waters are often referred to as 'marine deserts', with productivity generally less than about 30 g C m^{-2} yr^{-1}, which is comparable to a terrestrial desert.

In temperate open-ocean regions, primary productivity is linked closely to seasonal events. In the winter the sea surface cools and the thermocline breaks down, assisted by strong winds that mix the surface layers of the ocean. This allows surface waters to become well mixed with deeper, nutrient-rich seawater. However, light levels are low in the winter and limit primary production to low levels. In the spring, as the days lengthen and the sun gets higher in the sky, a time is reached when both light and nutrients become non-limiting and a spring bloom of primary production takes place. In the summer, although light is now abundant, a thermocline has become re-established as the surface waters warm, locking out the photic zone from the nutrient-rich deeper waters. Nutrients are now limiting and the spring bloom 'crashes'. In the autumn, the thermocline breaks down once again and nutrients are regenerated into the photic zone. If this occurs early enough in the autumn, while there is still sufficient sunlight, then nutrients and light both become non-limiting for a short time, and an autumn bloom may occur. This bloom will persist until light once again becomes a limiting factor in the late autumn and winter. Although occurring in a number of pulses, primary productivity in temperate oceans totals in the order of 70–120 g C m^{-2} yr^{-1}, similar to a temperate forest or grassland.

Some of the most productive marine environments occur in the coastal ocean above the continental shelves. This is the result of a phenomenon known as coastal upwelling which brings deep, cold, nutrient-rich seawater to the ocean surface, creating ideal conditions for primary productivity which can total more than $500 \text{ g C m}^{-2} \text{ yr}^{-1}$, comparable to a terrestrial rainforest or cultivated farmland. These hotspots of marine productivity are created by wind acting in concert with the planet's rotation.

The phenomenon can be explained in basic terms as follows. When a steady wind blows over the surface of the ocean it sets the surface layer moving in the direction of the wind. This top layer of moving seawater will in turn set another layer of seawater beneath it moving, although a bit more slowly than the layer above it, and so on down the water column until all of the wind's energy has been transferred into moving water. However, due to the planet's rotation, which creates a phenomenon known as the Coriolis Effect, each of these moving masses of seawater curls slightly to the right in the northern hemisphere and to the left in the southern hemisphere. This creates a characteristic pattern of water movement called the Ekman Spiral. The net outcome of an Ekman Spiral is that the average direction of flow of the seawater set in motion by the wind is at roughly right angles to the direction of the surface wind—to the right of the wind direction in the northern hemisphere, and to the left in the southern hemisphere. This net movement of water to the right or left of the wind direction is known as Ekman transport.

Coastal upwelling can occur when prevailing winds move in a direction roughly parallel to the edge of a continent so as to create offshore Ekman transport. Coastal upwelling is particularly prevalent along the west coasts of continents. In the southern hemisphere, when a steady wind blows roughly from the south along the western edge of a continent, the net movement of the top 100 metres or so of seawater is roughly westward away from the coast as a result of Ekman transport (see Figure 9). This mass

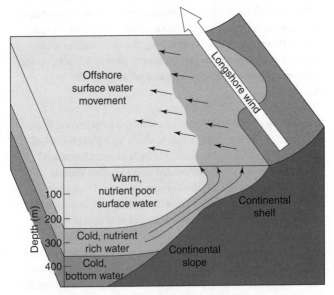

Offshore
surface water
movement

Longshore wind

Warm,
nutrient poor
surface water

Continental
shelf

Depth (m)

100
200
300
400

Cold, nutrient
rich water

Cold,
bottom water

Continental
slope

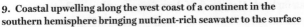

9. **Coastal upwelling along the west coast of a continent in the southern hemisphere bringing nutrient-rich seawater to the surface**

of displaced seawater can only be replaced by seawater from below, which is drawn up to the surface. If the upwelling is coming from water beneath the depth of the thermocline, then this replacement seawater is cold and nutrient rich. Likewise, in the northern hemisphere, when a steady wind blows from the north along the western edge of a continent, the surface layer of seawater moves on average westward away from the shore, creating coastal upwelling conditions. Since coastal upwelling is dependent on favourable winds, it tends to be a seasonal or intermittent phenomenon and the strength of upwelling will depend on the strength of the winds.

Important coastal upwelling zones around the world include the coasts of California, Oregon, northwest Africa, and western India in the northern hemisphere; and the coasts of Chile, Peru, and southwest Africa in the southern hemisphere. These regions are amongst the most productive marine ecosystems on the planet. When upwelling is occurring, the cold, nutrient-rich seawater stimulates immense blooms of phytoplankton, mainly the larger types such as diatoms and large species of dinoflagellates.

Considering the Global Ocean as a whole, it is estimated that total marine primary production is about 50 billion tonnes of carbon per year. In comparison, the total production of land plants, which can also be estimated using satellite data, is estimated at around 52 billion tonnes per year. Thus, the total primary production of the planet is a little over 100 billion tonnes of carbon per year, of which about 50 per cent comes from the oceans. However, the land and ocean systems generate this primary production in quite different ways. Primary production in the oceans is spread out over a much larger surface area and so the average productivity per unit of surface area is much smaller than on land. Also, most of the primary production in the oceans is generated by short-lived, rapidly reproducing microscopic organisms as opposed to the highly visible and often very long-lived trees, shrubs, and grasses in the terrestrial system.

El Niño Southern Oscillation

The upwelling off the west coast of South America supports one of the most productive fisheries on the planet—the Peruvian anchoveta fishery. Peruvian fishers have long been aware of a phenomenon they termed the El Niño (the Child) in honour of the Christ child, because it often occurred around Christmas. During El Niños the surface waters become unusually warm and fish and seabirds die, the anchoveta fishery diminishes or collapses, and whales, dolphins, and seals disappear.

We now know that this is part of an important and poorly understood global phenomenon termed the El Niño Southern

Oscillation (ENSO). ENSOs are associated with a reversal, or oscillation, of atmospheric pressure in the Pacific Ocean. Normally there is a persistent high-pressure system over Easter Island in the eastern Pacific Ocean, and a persistent low-pressure system over Indonesia in the western Pacific Ocean. Under these conditions the Pacific trade winds blow strongly from east to west, creating the coastal upwelling system that supports the Peruvian anchoveta fishery. However, during an El Niño event this pressure system reverses for some unknown reason, and the trade winds diminish, allowing warm water from the western Pacific Ocean to move eastwards and accumulate along the coast of South America, depressing the upwelling system that supports the anchoveta fishery. Without anchoveta, seabirds die and marine mammals that normally feed on anchoveta leave their usual feeding grounds. El Niños normally end in less than a year, although severe El Niños can persist for several years. Severe El Niño events have occurred in 1972–3, 1976, 1982–3, and 1997–8. The effects of severe El Niños are felt globally. California and western South America can experience unusually heavy rains, while parts of Australia, Indonesia, and Africa can experience severe drought. The flooding, crop failures, landslides, and other events associated with these abnormal weather patterns are very costly, causing huge damage and many deaths.

Moving energy through the marine system

Larger primary producers, such as diatoms and dinoflagellates, form the basis of the classic grazing food chain in the oceans. These phytoplankton cells are consumed by larger herbivorous grazers, in particular the ubiquitous copepods, an important group of zooplanktonic crustaceans. Copepods, which measure 1–2 mm in length, consume phytoplankton very efficiently by using their anterior appendages to create a stream of seawater that flows past their mouths where tiny hairs on their mouth appendages filter out the phytoplankton cells (see Figure 10). Zooplankton is then consumed by small fish and other marine

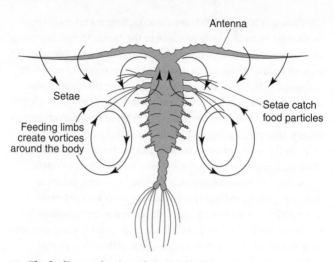

Antenna

Setae

Feeding limbs
create vortices
around the body

Setae catch
food particles

10. **The feeding mechanism of a copepod**

organisms, such as jellyfish, which are then consumed by a range
of larger predatory marine animals such as large fish, marine
mammals, marine turtles, and seabirds.

In highly productive coastal upwelling zones, zooplankton may
not always be an important link in the food chain. In these
systems small fish can often feed directly on large phytoplankton
cells and the fish are then eaten by larger fish and seabirds. This
makes for a very short and efficient food chain and one that is
exploited fully by humans who harvest the small fish in vast
quantities from these productive marine systems.

However, as mentioned earlier, a sizeable portion of primary
production in the oceans is produced by minute, free-floating
photosynthetic bacteria known as bacterioplankton. Because of its
small size, bacterioplankton is literally engulfed, or phagocytosed,
by other small (around 100 μm) single-celled organisms,
collectively called flagellates and ciliates, which occur at densities

of about 1,000 per millilitre of seawater. This step represents a major transfer of energy in the oceans. The flagellates and ciliates are then consumed by zooplankters and thus enter the grazing food chain via this route.

Much of the energy produced by photosynthetic marine organisms, perhaps about a quarter, is not actually transferred to the next trophic level through direct consumption of the organisms themselves. The oceans contain large amounts of dissolved organic matter (DOM) that has been leaked into the seawater from the cells of photosynthetic organisms. Viruses appear to be responsible for much of this leakage. Viruses are by far the most abundant 'life forms' in the oceans, existing at astonishing densities of around ten million per millilitre of seawater. Viruses must invade a host organism in order to survive and reproduce, and bacterioplankton and phytoplankton represent readily available hosts. These infections appear to be responsible for the death and break-up of the host cells and thus the leakage of large amounts of DOM into seawater.

This dissolved organic carbon is then absorbed directly by non-photosynthetic heterotrophic bacteria that use it as an energy source. These bacteria are in turn consumed by other microorganisms such as flagellates and ciliates, which can then be captured and eaten by larger zooplankters such as copepods. In this way, DOM, which was once considered to be a wasted source of energy in the oceans, is recycled back into the grazing food chain. This pathway of energy flow is often referred to as the microbial loop.

The huge mass of tiny faecal pellets produced by zooplankton, together with the disintegrated remains of dead phytoplankton and zooplankton, creates an immense store of particulate organic matter (POM) in the oceans, which forms the basis of yet another important pathway of energy flow—the detritus food chain. Some of this POM is consumed by zooplankton and hence gets recycled

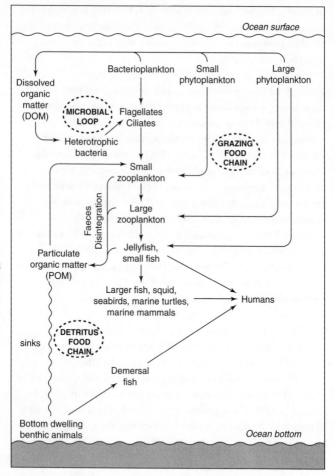

11. Major pathways of energy flow through the marine system

back into the grazing food chain. The remainder sinks to the
ocean bottom where it provides a food source for bottom-dwelling,
or benthic, organisms, which can then be consumed by bottom-
dwelling fish and other predators.

In summary, the energy of primary production in the oceans flows to higher trophic levels through several different pathways of various lengths (see Figure 11). Some energy is lost along each step of the pathway—on average the efficiency of energy transfer from one trophic level to the next is about 10 per cent. Hence, shorter pathways are more efficient. Via these pathways, energy ultimately gets transferred to large marine consumers such as large fish, marine mammals, marine turtles, and seabirds. These are all targeted in the end by a distinctly non-marine species, humans, for their own needs.

Chapter 3
Life in the coastal ocean

The coastal regions of the Global Ocean comprise a narrow strip of sea extending from the shoreline to the edge of the continental shelf. Although this coastal ocean environment is comparatively small, accounting for only about 7 per cent of the area of the Global Ocean, it is of huge importance to human society.

At this time, roughly 60 per cent of the human population, or over four billion people, crowd along the coast or live within 100 kilometres of a coast. This number is increasing rapidly as more and more people migrate to urban centres near coastal regions. It has been forecast that by the end of the century thirteen out of fifteen of the world's largest cities will be located on or near a coast.

This close relationship with human society means that the coastal ocean is heavily impacted by human activities. It is the receiving environment for many of the by-products of human society such as industrial and agricultural pollutants, human and animal sewage, and oil spills. The coastal ocean is heavily fished, providing much of the wild-caught seafood that humans obtain from the oceans. It is also the place where most aquaculture operations are sited, providing additional seafood for human consumption, but also leaving behind aquaculture waste products, such as unconsumed feed and faecal material, which accumulate and decompose on the seabed. Furthermore, the coastal ocean is heavily exploited for other vital resources such as oil and gas.

By virtue of its proximity to land, and its comparative shallowness, the coastal ocean is relatively easily studied by scuba diving, from smaller research vessels, and through the use of moored instruments that record information on a range of physical parameters, such as sea temperature, salinity, light levels, and currents. Thus, much information has been gathered about how coastal ecosystems function, and the biology of marine life in this region. If put to good use, this knowledge should, in theory, help us to better manage the coastal environment in the face of a rapidly growing human presence over the next four decades.

Kelp forest communities

Kelp forests are an important type of marine community found on rocky bottoms in shallow water close to shore. They are found mainly in cool, temperate coastal regions where sea temperatures normally do not exceed 20°C. They thrive particularly well in the cool, nutrient-rich seawaters of coastal upwelling zones (see Figure 12).

Kelp forests consist of dense aggregations of brown seaweeds, or kelps. Whereas most photosynthetic organisms in the oceans are microscopic and planktonic, kelps are large, multicellular algal forms that are attached firmly to the bottom by root-like structures known as holdfasts (see Figure 13). Although multicellular, kelps do not possess a specialized vascular system for transporting food and nutrients throughout the plant, and the holdfast is purely an attachment structure, not a root system. Instead, the kelp absorbs nutrients, water, and carbon dioxide directly from the surrounding seawater.

Kelps begin life as microscopic spores which settle to the seabed and grow into microscopic male or female plants, called

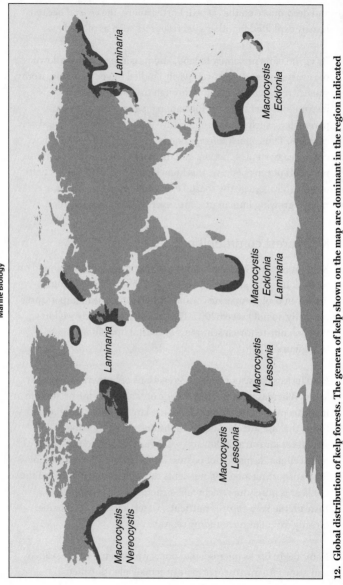

12. Global distribution of kelp forests. The genera of kelp shown on the map are dominant in the region indicated

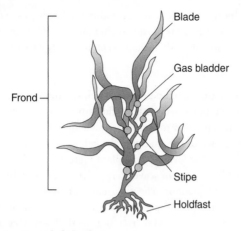

13. The structure of a kelp plant

gametophytes. The gametophyte is the sexual stage of the life cycle. The male gametophytes produce sperm, which are released into the sea where they fertilize eggs present on the surface of the female gametophytes. The fertilized eggs grow into the large plants, or sporophytes, that we generally associate with kelps. The sporophytes then release many new spores into the ocean to start the process over again. The life cycle takes around one to two years to complete.

Kelp forests are very productive. Under the right conditions kelps can grow very quickly, and some species reach a very large size. For example, the giant kelp, *Macrocystis*, which is abundant along the coasts of many parts of the world (see Figure 12), can grow at rates of more than 30 centimetres per day and reach lengths in excess of 30 metres in less than a year. The giant kelp forms vast underwater 'forests' with dense surface canopies that are held aloft in the seawater column with the assistance of gas-filled floats called pneumatocysts.

Kelp forests provide the foundation for a very lush, diverse marine community. Many types of organisms live attached to the surface of the kelps themselves, or buried within the interstices of the holdfasts, and the kelps provide shelter and food for many species of invertebrates and fish, including commercially harvested species.

Sea urchins are the main grazers of kelps, although normally sea urchins remove very little, perhaps 10 per cent or so, of the living kelp biomass directly, unless present in unusually large numbers. Much of the energy in the kelps enters the kelp forest community in the form of dead plant material which is consumed by a range of scavengers, including sea urchins and, where present, abalone. The sea urchins are fed on by a range of predators including species of sea stars, snails, octopus, spiny lobsters, crabs, and fish. In the North Pacific Ocean, sea otters also feed on sea urchins.

Longer-term studies of kelp communities have shown that they are unstable systems subject to rapid and substantial disruption. Kelp forests can be severely damaged by exposure to unusually warm seawater. This can result from climatic conditions that cause a reduction or cessation of upwelling of deep cool water along the coast. Kelp forests are also vulnerable to large storm-generated waves that can shred the kelps to pieces and tear many of them from their holdfasts. Such storms can be associated with severe El Niño events. The sea urchins remaining in the area then graze down the young kelps attempting to recolonize the area, preventing the re-establishment of a kelp forest and creating so-called urchin barren grounds. As the name suggests, these barren grounds are drab environments dominated by sea urchins and with low productivity and none of the species richness characteristic of kelp forests (see Figures 14a and 14b).

Urchin barrens can persist for long periods, and it often takes another catastrophic event to trigger a recovery of kelp forest to the area. This can take the form of a massive mortality of the

14a. Typical kelp forest

14b. A sea urchin-dominated barren ground that has replaced a kelp forest

sea urchin population, which reduces grazing pressure on young kelp plants, and clears the way for a kelp forest to be re-established. Such urchin mass mortality can be the result of another severe storm that this time devastates the sea urchins, or from an urchin disease, the spread of which can be encouraged by a period of unusually warm seawater that weakens the sea urchins and makes them more susceptible to the pathogen.

Changes in the abundance of key predators as a result of human interference can also severely impact kelp forests. Sea otters were once an abundant member of kelp communities in the coastal waters along the rim of the northern Pacific Ocean from northern Japan to Baja California. Their total population was estimated in the hundreds of thousands. Sea otters feed heavily on sea urchins, along with other marine invertebrates, such as abalone and crabs. By feeding on sea urchins, the sea otters reduce the intensity of grazing on the kelps, and help maintain stable kelp forest communities. If sea otters are missing from the system, sea urchin numbers can increase greatly, putting greater grazing pressure on kelp forests, and sometimes transforming them into urchin barren grounds. For this reason sea otters are often referred to as a keystone species, which is one that may not be present in large numbers but plays a fundamental role in shaping and maintaining the structure of a marine community.

There is evidence from the study of middens that aboriginal Aleut fishers in the Aleutian Islands hunted sea otters heavily enough to cause a shift of local coastal communities, from kelp dominated to sea urchin dominated, over 2,000 years ago. Commencing in the 1700s, sea otters began to be harvested on an industrial scale by fur traders for their dense pelts, and by the early 1900s only several thousand sea otters remained throughout their entire range, inhabiting only a few isolated coastal refugia. This resulted in the creation of extensive barren grounds in places like the

Aleutian Islands and off the coasts of Alaska, Canada, and elsewhere.

From the early 1900s, most commercial hunting of sea otters had been banned and conservation efforts had begun, with sea otters from surviving populations being reintroduced into certain areas. As a result of these efforts, sea otter populations have rebounded to about 100,000 animals occupying two-thirds of their former range, and a switch back to a kelp-dominated state has been observed in areas where sea otters are now abundant.

Kelp forests off California have been harvested by humans since the early 1900s. Generally only the upper metre or so of the kelp is removed in swathes from boats employing giant 'hedge clippers'. This practice appears to do little damage to the kelp community. In the past the kelp was used for the production of potash used in the manufacture of explosives and fertilizers. Today kelp is the principal source of algin, which is used extensively as an additive to thicken and stabilize a wide range of food products.

Seagrass meadows

Seagrasses are the basis of another important type of marine community that is widespread on sandy and mud bottoms in shallow temperate, subtropical, and tropical waters. Unlike kelps, seagrasses are flowering plants that have become adapted to live completely submerged in seawater. They evolved originally on land, and are the only group of flowering plants that have recolonized the marine environment.

Seagrasses have a true root and vascular system, and absorb their nutrients from the sediment on which they live (see Figure 15). The leaves are long, thin, flexible blades, generally about 10–50 centimetres in length, but up to a metre in length in some species. Seagrasses colonize the bottom by sending out underground stems, or rhizomes, which give rise to new plants. They can also

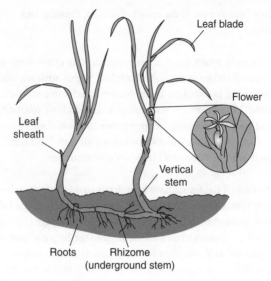

15. Structure of seagrass

reproduce sexually, with currents carrying pollen from one flower to another, and with the seeds also being dispersed by currents.

Seagrasses can cover extensive areas of the seabed and are very productive. The blades can grow as much as a centimetre per day and are continually being shed and replaced. Seagrasses can thus form vast, lush meadows comprising thousands of leaves per square metre of bottom. Seagrass meadows are most abundant in water less than about 10 metres deep, but can be found down to depths of 40 metres or so in very clear seawater. In temperate regions, *Zostera*, or eelgrass, is a widespread type of seagrass; while *Thalassia*, or turtle grass, is common in tropical regions.

Seagrasses form the basis of a complex community. Many different kinds of clinging and encrusting organisms colonize the surface of the seagrass blades, while various kinds of burrowing animals, such as clams and worms, are found among the roots. These organisms provide a rich source of food for fish and other animals, including commercially important species.

In temperate regions there are few animals that can graze directly on the tough seagrass blades, with the exception of some birds, such as swans, geese, and ducks, which, if present in large numbers, can sometimes overgraze large areas. The dead leaves and roots of seagrasses are broken down by bacteria and fungi and fed upon by a range of detritus feeders such as bacteria, worms, crabs, brittle stars, and sea cucumbers. In tropical waters, sea urchins graze on seagrasses and, when present in large numbers, can denude extensive areas of seagrass.

The green sea turtle, *Chelonia mydas*, also feeds on seagrass leaves, hence the common name, turtle grass, for some tropical species of seagrass. Green turtles mate at sea and the females return to the same beaches where they hatched to lay their own eggs. A female digs a pit on a dry area of the beach into which she lays her eggs. She then covers the eggs with sand and returns to the sea. A female may return several times to the same beach during the breeding season at intervals of several weeks to lay successive clutches of eggs. The eggs hatch after a couple of months, generally at night, and the hatchlings scuttle quickly to the sea. Here they remain in coastal waters near where they hatched to feed and grow. When they mature, the turtles migrate to new seagrass feeding grounds where they remain until they are ready to return to their nesting beaches to reproduce. These migrations can cover vast distances. For instance, one population of green sea turtles feeds on seagrasses along the Brazilian coast, but migrates over 2,300 kilometres to nesting beaches on tiny Ascension Island in the middle of the South Atlantic Ocean to breed. The turtles use the angle of the sun, wave direction, and smell to navigate their way to the island.

Worldwide, green turtles have been heavily hunted for their meat, and their eggs are harvested from their nesting beaches for food. As a consequence, green turtle numbers have declined drastically. For instance, it has been estimated that in the 17th century, somewhere between fifty million and a hundred million green turtles inhabited the Caribbean Sea, but numbers are now down to about 300,000. Since their numbers are now so low, their impact on seagrass communities is currently small, but in the past, green turtles would have been extraordinarily abundant grazers of seagrasses. It appears that in the past, green turtles thinned out seagrass beds, thereby reducing direct competition among different species of seagrass and allowing several species of seagrass to coexist. Without green turtles in the system, seagrass beds are generally overgrown monocultures of one dominant species.

Manatees and dugongs are large marine mammals that also forage on seagrasses. Dugongs live in the warm waters of the Indian and Pacific Oceans, whereas manatees are found in the Caribbean and Gulf of Mexico, and off the coast of West Africa. Dugongs in Australia have been shown to engineer seagrass beds in a way similar to green turtles. Herds of grazing dugongs thin out seagrass beds, creating space for more species of seagrass to coexist, and for younger, more nutritious plants to regenerate. This has been likened to dugongs 'cultivating' the seagrass meadows in a way that provides them with improved nutrition.

Seagrasses are of considerable importance to human society. Some coastal communities use the seeds and rhizomes of seagrasses as food, and seagrass is used as a packing material and a local source of fertilizer. Because seagrasses are effective at trapping and binding sediment, they stabilize the sea bottom and help protect coastlines from erosion. Seagrass beds also act as nursery grounds and foraging areas for many commercially important species of fish, as well as commercially important invertebrates such as clams, crabs, shrimp, and oysters.

It is therefore of great concern that seagrass meadows are in serious decline globally. In 2003 it was estimated that 15 per cent of the planet's existing seagrass beds had disappeared in the preceding ten years. Much of this is the result of increasing levels of coastal development and dredging of the seabed, activities which release excessive amounts of sediment into coastal waters which smother seagrasses. Seagrass meadows are also routinely torn up by boat propellers and anchors.

It is obvious that humans need to value seagrass habitats more highly and take the necessary steps to preserve them. This means protecting important seagrass habitats from coastal development and educating recreational boaters to not drive across shallow stretches of seagrass, and to anchor away from seagrass meadows. In some regions seagrass-friendly mooring systems are being promoted that minimize the impact on the bottom.

Soft-bottom communities

Vast expanses of the coastal regions of the Global Ocean consist of sand or mud bottoms. These habitats are strongly influenced by currents and wave action, especially in the shallower depths. Sandy bottoms are present where there are reasonably strong currents that carry away the fine, easily suspended mud particles, leaving behind the coarser sand grains. Muddy bottoms occur in areas with little current and tend to be rich in fine particles which settle out and collect in such areas. In either case, there is almost a complete lack of vegetation, and animal life dominates these habitats.

Most of the animals living in soft-bottom habitats are found buried in the sediment and are referred to as infaunal organisms. For the most part these comprise various species of burrowing clams and polychaete worms. Many of these are suspension feeders which filter plankton and small dead organic particles from the overlying seawater. Suspension-feeding clams extend long siphons from their burrows up to the surface of the sediment and pump in this suspended food material. The suspension-feeding worms often dwell in tubes from which they extend a set of tentacles used to filter out suspended particles. On the other hand, some of the infaunal clams and worms are deposit feeders, obtaining nutrition by directly consuming sediment and digesting the organic material and bacteria present in the sediment as it passes through their guts.

Not all members of the soft-bottom communities are burrowers, however. Some, the so-called epifaunal organisms, live on the surface of the sediment and include brittle stars, sea urchins, sea stars, sea cucumbers, sand dollars, snails, crabs, and shrimp. Many of these surface dwellers feed by either picking up organic particles lying on the surface of the sediment or, like infaunal deposit feeders, engulfing sediment and digesting the organic particles it contains. Others, such as snails, crabs, and sea stars are predators, feeding on other members of the community.

The epifaunal and infaunal members of soft-bottom communities support a range of bottom-feeding predatory fish that live on or near the seabed. These include skates and rays and many types of commercially important species such as haddock, pollock, hake, and cod, as well as flatfish such as flounder, halibut, and sole. These bottom-dwelling fish, often referred to as demersal fish or groundfish, are generally caught using bottom trawls, and play a vital role in feeding the human population.

Coastal dead zones

Many coastal marine systems are being greatly stressed and altered by the release of excess nutrients into coastal waters as a result of human activities, a process referred to as eutrophication. The two main culprits are nitrogen- and phosphorus-containing compounds.

The nitrogen cycle is one of the most human-altered nutrient cycles on the planet. The use of nitrogen-based fertilizers on land to maintain soil fertility and boost crop and pasture yields is a major source of nitrogen pollution. Some of the nitrogen compounds spread on agricultural land are not incorporated into crops but are leached by rain into streams and rivers and, hence, into the oceans. Also, in pastoral agricultural systems, nitrogen fertilizers are often added to boost pasture growth which is grazed by livestock. Livestock urine patches are extremely high in nitrogen, much of which leaches into streams and rivers after rainfall, and then into the oceans. Much of this nitrogen leaching from cropland and pastureland is in the form of nitrate. Other sources of nitrogen pollution are human wastes, as well as nitrogen produced during the burning of fossil fuels and organic biomass, such as wood and crop waste; this ends up in rivers and the oceans as nitric acid in acid rain.

The extensive use of phosphate-based fertilizers is a major source of phosphorus pollution. Some of the phosphate in the fertilizers that are spread on agricultural land is washed off into streams and rivers, mostly adhered to soil particles; from there it finds its way into the oceans. Human and animal sewage waste is another source of phosphorus compounds, as is deforestation. In the latter case, some of the soil exposed following the removal of trees, as well as the ash from burnt trees, both of which contain phosphorus, is washed into rivers. It is estimated that phosphorus concentrations in many rivers are now on average twice natural levels, and much of this ends up in the oceans.

Eutrophication of the oceans through excess nitrogen and phosphorus stimulates massive blooms of photosynthetic bacteria and plankton, particularly in coastal waters. As this mass of primary producers dies, it decays and consumes oxygen in the seawater. If oxygen levels are reduced below what is required to sustain most of the marine species in the area, then temporary or permanent marine 'dead zones' are the result.

The number of marine dead zones in the Global Ocean has roughly doubled every decade since the 1960s and now sits at over

500, occupying a portion of the ocean equal to that of the United Kingdom. Not surprisingly, dead zones are particularly common in areas subject to run-off from intensive agriculture.

Loss of biodiversity, high fish mortality, and the collapse of local fisheries are all associated with such dead zones. The Louisiana dead zone in the Gulf of Mexico is one of the largest known dead zones, covering an area of over 22,000 square kilometres, and it is still expanding. It is a seasonal feature that results from the huge amount of nutrients discharged, mainly in the spring, into the Gulf of Mexico from the Mississippi River, which drains a large area of intensively farmed agricultural land. Apart from the overall destruction of a natural marine system, this dead zone is impacting negatively on commercial and recreational shrimp and oyster fisheries in the Gulf. Other dead zones include parts of the Baltic Sea, the northern Adriatic Sea, and Chesapeake Bay.

Harmful algal blooms

Eutrophication is also contributing, at least in part, to more frequent and widespread occurrences of toxic algal blooms in coastal waters. These so-called harmful algal blooms (HABs) can reach such high cell densities that they discolour the sea, sometimes to a reddish hue, hence the common name red tides for this phenomenon.

HABs are created by a small number of phytoplankton species, often dinoflagellates, which produce a range of potent toxins that are present in their cells and can be excreted into the seawater. These toxins are then transferred through the food web, accumulating first in zooplankton that feed on the toxic phytoplankton and also in animals such as clams, mussels, scallops, and oysters that filter-feed on the toxic plankton. The toxins can then be transferred further up the food chain into fish, marine birds, and marine mammals. All of these organisms can be affected to a greater or lesser extent through the consumption of these toxins, sometimes resulting in massive mortalities of fish and other sea life, including sea birds and marine mammals, and causing the closure of local fisheries.

Humans feeding on contaminated shellfish and fish can experience neurological symptoms, such as tingling of the fingers and muscular paralysis, as well as respiratory problems and gastrointestinal symptoms such as diarrhoea, vomiting, and abdominal cramps. Such symptoms can be very severe and sometimes fatal. Toxic algae can sometimes become suspended in the air by wave action to form a coastal spray which, when inhaled by people, can cause asthma-like symptoms.

Occasional HAB events are a natural phenomenon and have probably always been a feature of coastal waters. Early explorers of the 17th and 18th centuries described occurrences of discoloured water and noted that indigenous coastal peoples would avoid harvesting shellfish at certain times of year in certain places for fear of being poisoned. But in recent decades HAB events have been observed occurring with greater frequency, becoming more persistent and affecting larger geographic areas. For instance, HAB events have become very common along the coasts of China since the 1970s, resulting in millions of dollars of damage yearly.

It is not entirely clear why this is occurring. It may in part be explained by improved recognition and reporting of toxic bloom and seafood poisoning events, but there is almost certainly a link to increased eutrophication of coastal waters that promotes algal blooms in general. Also, toxic algal species are being routinely transferred from one port to another around the world in the ballast water of ships. Ballast water is seawater pumped into the ballast tanks and cargo holds of ships to give them better stability when on a voyage. Ballast water is usually taken on when a ship has delivered cargo to a port and is leaving with less cargo or no cargo. Millions of litres of seawater are taken on at a time and then often transported and released at the next port where the ship picks up more cargo. This is a likely mechanism for expanding the range of toxic algal species. Some species have long-lived cyst stages that can lie dormant in the bottom sediments for many years until conditions are favourable for growth, and a HAB event is then triggered.

Biological invasions

Many other kinds of marine organisms in coastal waters besides toxic algae are pumped into the ballast tanks of ships. When a ship is in shallow water it can also pump in sediments and any associated bottom-dwelling organisms. When the ballast water is next released these organisms may also be released. In this way non-native, or exotic, invaders are introduced into areas where they would never normally be found without human intervention.

Roughly ten billion tonnes of ballast water are transferred globally each year and thousands of marine species are carried around the world in ballast water every day. Ships also move marine organisms long distances in other ways. Boring organisms, such as ship worms, can colonize the hulls of wooden ships, and attached organisms like barnacles and seaweeds can foul the hulls of ships. Such organisms can release their free-living larval stages in foreign ports that then settle to the bottom, allowing the organisms to colonize the new site.

Typically, very few of these foreign invaders will survive in their new surroundings. However, some encounter conditions that allow them to become well established and sometimes overwhelm the natural marine community in the area. This may be because the invader lacks the natural predators, pathogens, or parasites in its new location that would normally keep its numbers in check. Or it may encounter an unusually abundant food supply or is able to outcompete native species for available food and habitat space. Examples of introduced marine organisms are legion and include seaweeds, jellyfish, sponges, worms, crabs, barnacles, sea stars, clams, mussels, oysters, snails, fish, and many others.

The introduction of a jellyfish-like animal, the comb jelly, *Mnemiopsis leidyi*, from the coastal waters of North America into the Black Sea in ballast water in the 1980s illustrates very well the devastation that can be caused by marine invaders. This animal quickly multiplied to plague numbers in the predator-free environment of the Black Sea, voraciously consuming the natural zooplankton in the sea, including the eggs and juvenile stages of fish. Fish stocks collapsed by the early 1990s, causing great economic loss to the region, and dolphins, which fed on these fish, disappeared. Interestingly, it took the invasion of another exotic species of comb jelly, *Beroe ovata*, also in ballast water, to alleviate this ecological disaster in the Black Sea. Around 1997, this species began to thrive in the Black Sea, feeding heavily on the first foreign invader, causing a steep reduction in its numbers. The *Beroe ovata* population then collapsed as it exhausted its food supply. Since then, fish stocks have begun to recover and dolphins have returned.

The introduction of the Japanese sea star (*Asterias amurensis*) into Australia is another good example of the dramatic impact that an exotic species can have when it invades a new habitat. This sea star is native to the coastal waters of Japan, northern China, Korea, and Russia, but sometime in the 1980s it was introduced into Tasmania, probably as larvae in ballast water, or as juveniles clinging to the hulls of ships arriving from the North Pacific.

The population of this sea star exploded in its new location and by the mid 1990s had reached extraordinary densities in some places. For example, in the Derwent Estuary of Tasmania there are an estimated thirty million individuals at densities up to 10 per m^2. This sea star is a voracious predator feeding on just about anything in its path, including shellfish, crabs, sea urchins, sea squirts, and other sea stars, and turning the bottom into a virtual monoculture of foreign sea stars. They also pose a threat to aquaculture operations in the area with the potential to decimate mussels, oysters, and scallop farms.

Once well established, these foreign invaders are impossible to eradicate. Attempts have been made to control the spread of the Japanese sea star by recruiting divers to remove the animals by hand, and also by trapping or dredging the sea stars. They have also been commercially harvested and converted into fertilizer. But none of this has had much success in restoring the natural marine system of the area. Efforts are now focused on limiting the spread of the species through an education campaign that encourages reporting of local sightings, which is followed up by a rapid eradication programme. There is fear that the species could spread to New Zealand, and legislation has been passed in that country preventing the discharge in New Zealand ports of ballast water that has been taken from ports in Australia where this sea star is now found. This is to reduce the chance of introduction of the species into New Zealand in its planktonic larval stage.

International efforts are now being made to limit the spread of exotic marine species in ballast water. Ships are now meant to empty, and then recharge, their ballast tanks in the open ocean before arriving at a port. The reasoning behind this is that ballast water hitch-hikers taken up in port will be released into the open ocean where they cannot survive, and that the planktonic organisms taken up in the open ocean will be released into the coastal waters of the next port where conditions will not be suitable for survival. Unfortunately, not all ships follow this procedure, and ballast water cannot be safely exchanged in the open ocean during rough weather. Thus, some port authorities are considering developing procedures to sterilize a ship's ballast water before it is pumped out, or to pass the ballast water through a treatment facility on shore before its discharge into the natural environment.

Plastic debris

For the past sixty years or so, plastic materials, which are derived from oil and gas, have become a ubiquitous and indispensable product of human society, and a major source of human waste. Not surprisingly, huge amounts of this waste plastic end up in the marine environment. Most marine plastic debris—around 80 per cent—comes from the land. Large amounts of discarded plastic materials are discharged into the oceans, either directly or via rivers, from overflowing sewers, and from storm drains during heavy rainfall. Ships and boats are the other main source of marine plastic pollution—a great variety of plastic trash is routinely dumped overboard from commercial and recreational vessels, and fishing boats discard or lose large amounts of fishing gear, such as fishing lines and nets.

A major problem with plastic debris is its persistence. Plastic materials, prized by humans for their inertness and durability, degrade very slowly and will persist in the environment for hundreds or even thousands of years. The oceans have thus been subject to a half century or more of accumulated plastic waste, with no sign of abatement. Plastic debris is thus now common everywhere in the oceans—floating on the surface, accumulating on the seafloor at all depths, and littering all coasts. In the oceans some plastic materials, such as polystyrene, are broken up by wave action into smaller fragments, and can eventually become tiny (<5 mm) 'microplastic' fragments, sometimes called mermaid's tears, that accumulate in marine sediments or remain suspended in seawater.

The amount of plastic debris stranded on shorelines throughout the world is appalling. In the Caribbean there can be anywhere from 1,900 to over 11,000 separate items of significant plastic debris along a kilometre of shoreline. In Indonesia more than 29,000 plastic items have been recorded along a kilometre stretch of shoreline. Surveys of the seabed show that there are typically hundreds of items of plastic debris per square kilometre, and in some places in Indonesia and the Caribbean there can be thousands of items of plastic per square kilometre of seabed. Levels of floating plastic debris have been quantified using visual surveys from ships and show that floating plastic objects are present in significant quantities in all oceans and are particularly common in coastal waters. For example, anywhere from ten to more than a hundred pieces of floating plastic per square kilometre have been recorded in the English Channel. Enormous amounts of plastic are present in the oceanic gyre systems of the Global Ocean which, on account of their circular motion, tend to trap and accumulate floating debris. A study using nets to collect floating debris in the North Pacific gyre revealed an astounding average of 334,271 pieces of plastic per square kilometre. Not surprisingly, this system is often called the Great Pacific Garbage Patch.

Plastic debris is very harmful to many forms of marine life which become entangled in it or mistake it for food and ingest it. Hundreds of different species have been recorded being harmed by plastic debris in very large numbers, including seabirds such as penguins, albatrosses, pelicans, and many shorebirds; marine mammals, including whales, seals, sea lions, sea otters, manatees, and dugongs; and various species of sea turtle. Entanglement is often caused by discarded fishing nets and ropes, monofilament fishing line, packing strapping bands, and six-pack rings. Seabirds and sea turtles routinely ingest plastic debris of all sorts, probably mistaking it for items of food. Sea turtles, for example, appear to mistake plastic bags for jellyfish, one of their prey items. Ingestion of plastic can lead to obstruction of the gut and subsequent death, whilst toxic chemicals may leach out of the material and cause other harmful effects.

The scale of the plastic debris problem is daunting, but some measures have been put in place to help stem the enormous flow of waste plastic into the marine environment. A number of

international and regional conventions have been agreed that ban
the disposal of plastic at sea. However, enforcement is a problem
and millions of tonnes of plastic a year are still being dumped
from ships. Also, the majority of plastic debris comes from land,
not sea-based, sources. To meaningfully address this issue,
countries will have to adopt and enforce strategies that include
waste minimization and reuse and recycling of plastic materials.

Chapter 4
Polar marine biology

Flourishing marine biological systems are present in the extreme environments of the Arctic and Antarctic polar regions of the planet. Both these regions are characterized by constantly cold seawater temperatures, ice-covered seas, and extreme seasonal fluctuations in light levels. In many other ways, however, these regions are very dissimilar and have evolved strikingly different and unique marine ecosystems.

Marine biology of the Arctic Ocean

The Arctic Ocean is a comparatively small (14.6 million square kilometres) and rather isolated body of seawater with extensive areas of shallow, continental shelf. It is largely surrounded by land masses with only two outlets, the very narrow Bering Strait to the Pacific Ocean, which is only 70 metres deep; and the broader 400-metre-deep Fram Strait to the Atlantic Ocean. Several large Siberian and Canadian rivers empty into the Arctic Ocean, creating a thin, lower-salinity layer of seawater about 20–50 metres deep that floats on the saltier and denser seawater beneath. Extensive areas of the Arctic Ocean have a soft sedimentary bottom resulting from the discharge of large amounts of sediment from these river systems.

The surface of the Arctic Ocean is at or near the freezing point of seawater (–1.9°C) for much of the year. Thus, much of the Arctic Ocean is permanently covered by a floating cap of sea ice which expands and retreats with the seasons. The cap is largest in April at the end of the Arctic winter (about 15 million square kilometres on average), and smallest in September at the end of the Arctic summer (about 7 million square kilometres on average). The summer melt occurs mostly over the vast continental shelves of the Arctic Ocean, while most of the central Arctic Ocean remains covered by ice throughout the year. This 'multi-year' sea ice has survived complete melting for several years and is 3–4 metres thick. The remainder is thinner first-year ice about 1–2 metres thick.

One might expect this vast volume of frozen seawater to be devoid of life. In reality it harbours an abundant and diverse marine community, unique to polar seas, and one which plays a fundamental role in sustaining the polar food web. Sea ice is habitable because, unlike solid freshwater ice, it is a very porous substance. As sea ice forms, tiny spaces between the ice crystals become filled with a highly saline brine solution resistant to freezing. Through this process a three-dimensional network of brine channels and spaces, ranging from microscopic to several centimetres in size, is created within the sea ice. These channels are physically connected to the seawater beneath the ice and become colonized by a great variety of marine organisms.

A significant amount of the primary production in the Arctic Ocean, perhaps up to 50 per cent in those areas permanently covered by sea ice, takes place in the ice. In the long Arctic summers, sufficient light penetrates the snow covering the ice, and the ice itself, to sustain primary productivity by organisms in the ice. At the start of the summer, diatoms and photosynthetic dinoflagellates appear within the sea ice, and also in a layer coating the bottom of the ice. These photosynthetic organisms soon become so abundant that they colour the ice brown

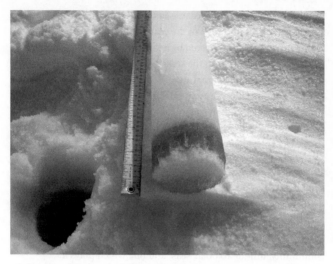

16. Ice core showing dark band of ice algae

(see Figure 16). The sea ice provides a uniquely stable marine habitat that keeps these photosynthetic organisms within the photic zone at all times during the summer, thus maximizing primary productivity.

Many other kinds of microorganisms inhabit the sea ice, including viruses, bacteria, fungi, protozoan ciliates, and flagellates, and a well-developed microbial food web is present similar to the more conventional microbial loop of the open-ocean pelagic zone. Photosynthetic organisms in the ice leak dissolved organic matter (DOM) into the brine channels, which is absorbed by bacteria as an energy source. These bacteria are then consumed by an abundant population of flagellates and ciliates.

Large numbers of zooplanktonic organisms, such as amphipods and copepods, swarm about on the under surface of the ice, grazing on the ice community at the ice–seawater interface, and sheltering in the brine channels. Organisms that would normally be found on the ocean bottom are also part of this sea-ice community, including flatworms and nematodes.

These under-ice organisms provide the link to higher trophic levels in the Arctic food web (see Figure 17). They are an important food source for fish such as Arctic cod and glacial cod that graze along the bottom of the ice. These fish are in turn fed on by squid, seals, and whales. The seals are an important source of food for the 25,000 or so polar bears which inhabit the Arctic region. Polar bears are adept at killing seals as they emerge through breathing holes in the ice, or when they haul themselves up onto the ice edge.

Partway through the summer, as the edge of the sea ice begins to melt, ice algae are released into the seawater beneath the ice and seed an under-ice phytoplankton bloom. As the summer progresses, and the ice edge breaks up and retreats, this bloom creates a 20–80-kilometre-wide zone of extremely high productivity at the ice edge. Walrus, seals, narwhal, beluga whales, and bowhead whales are abundant at this ocean–ice boundary, along with seabirds and polar bears. This community then follows this ice edge oasis for hundreds of kilometres as it retreats northwards during the Arctic summer.

Unconsumed organic material from these Arctic blooms sinks to the seabed and supports the Arctic benthic community. Amphipods, worms, clams, and brittle stars live on or within the soft-bottom sediments of the Arctic Ocean, and are particularly common on the continental shelves where they are fed on by bottom-feeding fish such as eelpouts and sculpins, as well as gray whales and walruses that forage on the seabed.

Some year-round areas of open water do persist in certain places in the Arctic ice cap. These surprising features, called polynyas, occur where ocean currents, strong winds, or warm water

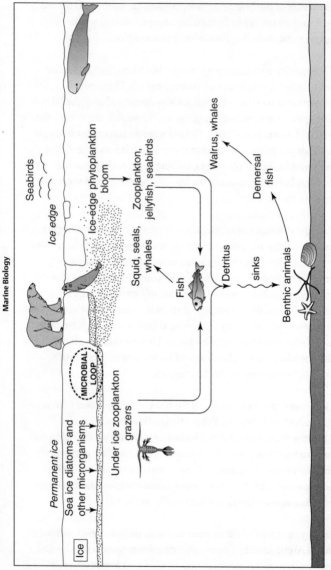

17. Depiction of the Arctic food web

upwellings prevent the surface waters from freezing. Most polynyas are only a few square kilometres in size, although some, such as the North Water Polynya, are thousands of square kilometres in area. These islands of water serve as year-round refugia for air-breathing mammals, including seals and whales, and some whales will reside within polynyas throughout the winter, forgoing their normal southward autumn migration ahead of the advancing ice cap. In the Arctic spring, polynyas become highly productive marine oases fuelled by the season's first light, whose penetration is unhindered by snow and ice cover. Large numbers of seals, polar bears, marine mammals, and seabirds congregate within and around these features.

As a result of human-induced climate change, the Arctic region is warming, and it is doing so at a rate faster than the rest of the planet. This is having a great impact on the Arctic Ocean ice cap. The maximum winter extent of sea ice in the Arctic Ocean has been decreasing by an average of 3 per cent per decade since 1979, and the overall thickness of the ice has decreased as well. The extent of minimum summer ice cover is decreasing at an even faster rate—in recent years the extent of sea-ice cover at the end of the Arctic summer has been well less than 6 million square kilometres versus the longer-term average of around 7 million square kilometres. At this rate, the Arctic Ocean will soon become nearly or completely ice free in the summer, probably sometime in the 2040s or sooner.

Given the importance of the sea-ice community in the Arctic marine food web, it is clear that the Arctic marine ecosystem will be disrupted in fundamental ways in the near future. Overall primary production may increase, since less snow and ice cover means a deeper photic zone. However, the disappearance of the unique sea-ice community will undoubtedly impact on the Arctic food web in ways difficult to predict at this time. Seals and polar bears, adapted to living in close association with sea ice, will be particularly hard hit, since their feeding and breeding habitat will be greatly reduced. There are no commercial fisheries in the Arctic Ocean at this time because of the difficulty of operating fishing vessels in ice-covered seas. However, this may change in future as the Arctic ice cap shrinks.

Marine biology of the Southern Ocean

The Arctic and Antarctic marine systems can be considered geographic opposites. In contrast to the largely landlocked Arctic Ocean, the Southern Ocean surrounds the Antarctic continental land mass and is in open contact with the Atlantic, Indian, and Pacific Oceans. Whereas the Arctic Ocean is strongly influenced by river inputs, the Antarctic continent has no rivers, and so hard-bottomed seabed is common in the Southern Ocean, and there is no low-saline surface layer, as in the Arctic Ocean. Also, in contrast to the Arctic Ocean with its shallow, broad continental shelves, the Antarctic continental shelf is very narrow and steep.

The approximate northern boundary of the Southern Ocean is often designated as the 60°S latitude. Since the edge of the Antarctic continent is at roughly 70°S, the Antarctic marine system consists of a ring of ocean about 10° of latitude wide—roughly 1,000 km.

The Antarctic continent is covered by a thick ice sheet that flows outwards from the centre of the continent towards the coasts, and then out into the ocean to form a vast, floating, 100-metre-thick mass of permanent ice called the ice shelf. Seaward of the ice shelf, the ocean freezes seasonally.

The seasonal fluctuations in the Southern Ocean are enormous. At the start of the southern hemisphere winter, the sea begins to freeze, the freezing front moving outwards rapidly at a rate of about 4 kilometres per day. By the end of the southern hemisphere winter, roughly 18 million square kilometres of ocean are covered by sea ice. Unlike in the Arctic Ocean, however, almost all of this sea ice will melt during the summer, with only about 3 million square kilometres remaining. Since most of the sea ice in the Southern Ocean is just one year old, it is comparatively thinner than in the Arctic Ocean, generally only 1–2 metres thick.

Antarctic waters are extremely nutrient rich, fertilized by a permanent upwelling of seawater that has its origins at the other end of the planet. As described in Chapter 1, cold dense seawater formed in the North Atlantic Ocean—the North Atlantic Deep Water—sinks and flows slowly southward near the bottom of the Atlantic basin, to emerge hundreds of years later off the coast of Antarctica. This continuous upwelling of cold, nutrient-rich seawater, in combination with the long Antarctic summer day length, creates ideal conditions for phytoplankton growth, which drives the productivity of the Antarctic marine system.

As in the Arctic, a well-developed sea-ice community is present. Antarctic ice algae are even more abundant and productive than in the Arctic Ocean because the sea ice is thinner, and there is thus more available light for photosynthesis. Diatoms can be particularly abundant, and as the sea ice breaks up and recedes in the spring, these are released into the open water at the ice edge and seed huge blooms of phytoplankton, often dominated by large diatoms.

This massive pulse of primary productivity supports Antarctica's most important marine species, the Antarctic krill, *Euphausia superba*. Antarctic krill are shrimp-like, nearly transparent, zooplanktonic animals about 4–6 centimetres in length (see Figure 18). They can live for five to ten years and thus are able to survive through successive long, dark Antarctic winters when phytoplankton is absent and other food is scarce. Krill are very adept at surviving many months under starvation conditions—in the laboratory they can endure more than 200 days without food. During the winter months they lower their metabolic rate, shrink in body size, and revert back to a juvenile state. When food once

18. Antarctic krill

again becomes abundant in the spring, they grow rapidly and redevelop sexual characteristics.

In the Antarctic spring, krill crawl about under the sea ice, scraping the ice and grazing on the lush lawn of Antarctic ice algae. As the sea ice breaks up they leave the ice and begin feeding directly on the huge blooms of free-living diatoms, using their filter-feeding appendages. With so much food available they grow and reproduce quickly, and start to swarm in large numbers, often at densities in excess of 10,000 individuals per cubic metre—dense enough to colour the seawater a reddish-brown.

Krill swarms are patchy and vary greatly in size, from some that can be quite small—just patches of a few square metres when looking down on the sea surface—to ones that are hundreds of square kilometres in size and contain millions of tonnes of krill. Because the Antarctic marine system covers a large area, krill numbers are enormous, estimated at about 600 billion animals on average, or 500 million tonnes of krill. This makes Antarctic krill one of the most abundant animal species on the planet in terms of total numbers and biomass—by comparison, the human population is around seven billion, with a total biomass roughly similar to that of Antarctic krill. It is not surprising, then, that krill play a pivotal role in the Antarctic marine system.

Antarctic krill are the main food source for many of Antarctica's large marine animals, and a key link in a very short and efficient food chain (see Figure 19). Krill comprise the staple diet of icefish, squid, baleen whales, leopard seals, fur seals, crabeater seals, penguins, and seabirds, including albatross. Thus, a very simple and efficient three-step food chain is in operation—diatoms eaten by krill in turn eaten by a suite of large consumers—which supports the large numbers of large marine animals living in the Southern Ocean.

Blue, right, and fin whales are abundant in Antarctic waters during the summer. These whales have sieve-like baleen plates in their mouths which they use to capture large quantities of krill. They take in large mouthfuls of krill-laden seawater and then use their tongue to force the seawater out through the baleen sieve, which retains the krill. Although baleen whales are all competing for the same food source, they have evolved ways to partition their food amongst themselves and avoid direct competition for food. Different species of whales have different baleen sieve sizes and they feed at different depths. In this way they are targeting different sizes of krill at different depths.

All Antarctic whale species are migratory, feeding in Antarctic waters during the southern hemisphere summer, and then swimming long distances to warmer northern waters to breed during the winter months. In the southern hemisphere spring they head back south, following the retreating ice edge.

Crabeater seals, despite their name, also depend on krill as a food source. These seals, which live on ice floes, have specialized teeth

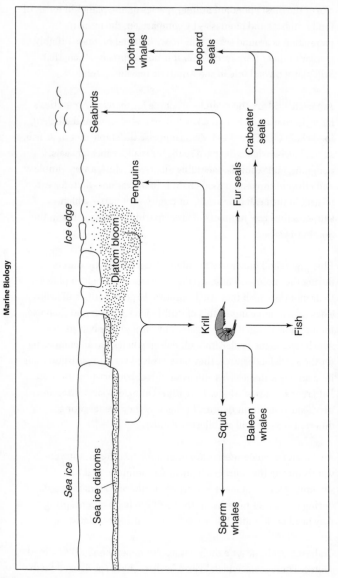

19. The central place of Antarctic krill in the Southern Ocean food web

that are adapted for straining krill directly from the sea. Much like baleen whales, they take a mouthful of seawater and expel it through their teeth, which retain the krill. Not surprisingly, given the huge size of their food stock, crabeater seals are very abundant; there are around fifty million of them in Antarctic waters, which makes them one of the most numerous large mammals on the planet, after humans.

Antarctic penguins also rely on krill. The most abundant species of penguin in Antarctica is the Adelie penguin. There are about 2.5 million breeding pairs, which feed on krill and small fish. They are capable of diving to depths of hundreds of metres in pursuit of food. The young penguins are especially dependent on krill, and if krill numbers are low in any particular season, juvenile mortality is high.

Large marine predators are common in the Antarctic marine system. Leopard seals are voracious carnivores well equipped to feed on large prey, such as penguins and crabeater seals. They too, however, can feed on the ubiquitous Antarctic krill and, like the crabeater seal, have some of their teeth modified to act as krill strainers. Killer whales, or orcas, are another common Antarctic predator, feeding on fish, penguins, seals, and other whales, but even orcas will feed on krill.

Many different species of squid are common in Antarctic waters and are a very important food source for large marine animals, including sperm whales and seabirds. One of the largest invertebrates on the planet, the colossal squid, inhabits the deep waters of Antarctica. Until recently, a complete specimen of this deep-sea animal had never been seen, and it was known only from bits found in the stomachs of sperm whales hunted down by whalers. However, in 2007, a live colossal squid was brought to the surface from a depth of around 2,000 metres in the trawl net of a deep-sea fishing vessel from New Zealand. This animal was 10 metres in length and weighed close to 500 kilograms. Stomach analysis shows that colossal squids are the major prey of sperm whales in Antarctic waters. Many sperm whales have scars on their body which appear to be caused by the sharp hooks present at the tips of the colossal squid's two long feeding tentacles, evidence of mighty predator–prey struggles in the cold abyssal depths of Antarctic waters.

An unusual group of fishes, the icefish, are common in Antarctica. These fish, which live constantly in seawater which is on the verge of freezing, have very little of the oxygen transporting red pigment, haemoglobin, in their blood streams. Oxygen is simply transported in solution in the blood plasma of these fish. These fish can obtain sufficient oxygen in this fashion because their body fluids are very cold and the amount of oxygen in solution increases with decreasing temperature. They also have other adaptations that prevent their bodies from freezing. For example, their body fluids contain complex proteins and sugars that provide a kind of antifreeze protection by lowering their freezing point.

Seabirds, including albatrosses, petrels, and fulmars, range widely in the Southern Ocean, feeding on krill, squid, and fish. The Wandering Albatross (*Diomedea exulans*) is perhaps the quintessential Southern Ocean seabird. It is the largest of the albatrosses, with a wing span of 3 metres or more, and spends most of its life on the wing, gliding on the Southern Ocean wind systems. Albatrosses will range over thousands of square kilometres of ocean foraging for food, and can travel up to 1,000 kilometres in a day. Albatrosses return to land to breed, mainly on subantarctic islands.

Despite the extreme cold, the sea floor communities of the Southern Ocean can be extraordinarily rich. In shallow waters less than about 15 metres, the seabed is routinely scoured by grounded ice. Here sessile marine animals are absent and the seabed is occupied during the ice-free period by mobile animals such as sea stars, sea urchins, and large nemertean worms that invade the area when possible to feed on diatoms growing on the sea floor and to scavenge on dead and dying animals.

In deeper water, attached benthic animals such as sea anemones, corals, and sponges proliferate in great numbers. In some areas, the densities of these sea floor animals are among the greatest recorded in any marine environment on the planet.

Antarctic bottom-dwelling animals in shallow water are subject to a curious phenomenon known as anchor ice mortality. Very cold seawater often freezes around animals living on the bottom and the resulting ice becomes buoyant enough to lift them from the sea floor. In some years anchor ice is capable of completely stripping the bottom of all large animals down to depths of 30 metres or so.

Southern Ocean marine mammals have been ruthlessly exploited in the past. Fur seal hunting began in the late 1700s, and by 1830 most fur seal colonies had been exterminated or reduced to a size where it was uneconomical to continue to hunt them. They were declared a protected species in 1964 and they now number more than four million individuals, which is probably a larger population than when exploitation first began.

Antarctic whaling commenced in the early 1900s, initially targeting humpback whales, and then spreading rapidly to other species, including southern right, blue, fin, sei, and sperm whales. In the years before the Second World War, tens of thousands of whales were taken annually and in the period between 1956 and 1965, 631,518 whales were recorded killed. The industry collapsed in the 1960s when it became uneconomical to hunt the remnant populations of whales that remained. By that time, the southern right and humpback whale populations had been reduced to about 3 per cent of their original populations, blue whales to about 5 per cent, and fin and sei whales to about 20 per cent. A moratorium on Antarctic commercial whaling came into force in 1986 and, apart from some 'scientific' whaling by Japan, no whales are now exploited.

Following the demise of seal and whale populations, the exploitation of Antarctic marine species moved down the food web to target smaller animals at lower trophic levels. Commercial fishing began in the late 1960s, first focused on species such as the mackerel icefish. In the 1980s, fishing vessels began to exploit the Patagonian toothfish using longlines set at depths of around 1,000 metres. This fish is marketed as 'Chilean sea bass' and has become a very popular fish commanding a premium price. Demand outstrips supply, which has lead to illegal fishing by vessels which have not been assigned a legal quota for this species. This longline fishery has recently expanded to a related species, the Antarctic toothfish. Antarctic toothfish are food for sperm whales, killer whales, Weddell seals, and large squid, so their removal will potentially impact on these dependent species.

Even the krill in the Southern Ocean are subject to human exploitation. Antarctic krill fishing began in the 1970s and by the early 1980s about half a million tonnes were being harvested annually. Catches then declined to around 100,000 tonnes per year as most nations abandoned the fishery because of the high cost of operating in the Southern Ocean. However, the demand for krill appears to be increasing again. Krill are being caught and processed into fish meal, which is a major component of the artificial feeds used on fish farms. Krill is also being used as a source of health food supplements, such as omega-3 oils.

This long-standing exploitation of Antarctic marine resources belies the commonly held notion that Antarctica possesses one of the planet's last pristine marine systems. Human impacts have undoubtedly pushed the Antarctic marine system out of its natural equilibrium, although it has been difficult to document the changes because of Antarctica's remoteness, the lack of an historical baseline of what is 'natural' in the Southern Ocean marine system, and the complexity of species interactions.

The abundant whale populations present prior to human exploitation would have consumed a significant proportion of the available krill in the Antarctic system. As whales were exterminated, 'surplus' krill that would have normally been eaten by the whales became available to other species such as seals, squid, and seabirds, likely allowing their populations to increase above equilibrium levels. This may explain why fur seal populations have recovered beyond their pre-exploitation levels. This situation may also contribute to a very slow recovery of whale populations because, in addition to whales having very low rates of reproduction, whales now have to compete for their krill food source with other species which have taken over their food niche.

A major human impact of another kind further complicates the story. It appears that human-induced global warming is resulting in a decline in Antarctic krill biomass in the very productive southwest Atlantic sector of the Southern Ocean, perhaps by as much as 80 per cent, since the 1970s. This appears to be the result of rising temperatures in Antarctica causing shrinkage in the Antarctic sea ice in this part of the Southern Ocean. Krill rely on the ice for their ice-algae food during the winter months and for shelter from predators, and it has been observed that in years when the extent of sea ice is low, krill are less abundant in subsequent years. It also appears that as krill stocks decrease in this region, salp numbers are increasing. Salps are planktonic, gelatinous animals that can live in warmer, less productive waters than krill. Thus the recovery of whale populations in the Southern Ocean may be further hindered by lower krill densities.

The depletion of the planet's stratospheric ozone layer caused by the release of synthetic chemicals, principally chlorofluorocarbons (CFCs), is also impacting on the Southern Ocean. Ozone depletion is very dramatic over Antarctica, particularly in the southern hemisphere spring between September and December when the ozone layer is thinned by about 50 per cent. This creates an ozone 'hole' allowing increasing levels of high-energy ultraviolet-B (UV-B) radiation to penetrate the surface waters of the Southern Ocean to about a depth of 2 metres. Overexposure to UV-B radiation impairs photosynthesis of plants, including phytoplankton, and studies have shown that the ozone hole results in at least a 6–12 per cent reduction in phytoplankton primary productivity in coastal areas of the Southern Ocean. The impacts of this on the ecology of the Antarctic marine system are unclear at this stage. However, given that further reduction in average global stratospheric ozone levels is predicted over the next century, despite international efforts to control the production and use of CFCs, it is important that efforts continue to better understand the effects of increasing UV-B radiation levels on Southern Ocean food webs.

Chapter 5
Marine life in the tropics

The tropical marine environment encompasses those parts of the Global Ocean where the surface waters are consistently warm throughout the year, rarely falling below 20°C. Such regions occur within an oceanic belt straddling the equator from roughly the Tropic of Cancer in the northern hemisphere to the Tropic of Capricorn in the southern hemisphere (23°N latitude to 23°S latitude).

Coral reefs

Coral reefs embody the iconic image of a tropical marine environment and are globally significant natural systems in terms of their beauty, their biological diversity, their productivity, and their economic significance (see Figure 20). These 'rainforests of the ocean' are very complex systems that are home to an incredible diversity of marine organisms—about a quarter of all marine species—perhaps upwards of two million different species of plants and animals. Coral reef systems provide food for hundreds of millions of people, with about 10 per cent of all fish consumed globally caught on coral reefs. They serve as natural protective barriers, sheltering coastal communities from the waves generated by hurricanes and typhoons. They are also the basis of employment through tourism for millions of people in the many countries with reefs in their coastal waters.

20. Aerial view of coral reefs of Heron Island, Great Barrier Reef, Australia

Physical requirements

Notwithstanding their importance, coral reefs occupy a very small proportion of the planet's surface—about 284,000 square kilometres—roughly equivalent to the size of Italy. This is because the physical requirements of the main reef-building animals—the corals—are very specific.

Reef-building corals thrive best at sea temperatures above about 23°C and few exist where sea temperatures fall below 18°C for significant periods of time. Thus coral reefs are absent at tropical latitudes where upwelling of cold seawater occurs, such as the west coasts of South America and Africa. Corals also require lots of light to thrive, so they are generally restricted to areas of clear water less than about 50 metres deep.

Reef-building corals are very intolerant of any freshening of seawater below a salinity of about 30 and so do not occur in areas exposed to intermittent influxes of freshwater, such as near the mouths of rivers, or in areas where there are high amounts of rainfall run-off. This is why coral reefs are absent along much of the tropical Atlantic coast of South America, which is exposed to freshwater discharge from the Amazon and Orinoco Rivers.

Finally, reef-building corals flourish best in areas with moderate to high wave action, which keeps the seawater well aerated, brings in a constant supply of food for the corals, and removes light-blocking sediment from the surface of the corals.

Spectacular and productive coral reef systems have developed in those parts of the Global Ocean where this special combination of physical conditions converges, such as in the Caribbean Sea; the many islands of Indonesia, the Philippines, the South Pacific, and the tropical Indian Ocean; in the Red Sea; and off the northeast and northwest coasts of Australia (see Figure 21).

Biology of corals

Reef-building corals, also known as scleractinian or stony corals, are colonial animals related to sea anemones. Each colony consists of thousands of individual animals called polyps (see Figure 22). Colonies grow through asexual reproduction—the polyps repeatedly bud off new polyps, creating an expanding layer of genetically identical polyps which share a common stomach cavity. As the colony grows, the polyps extract calcium from the surrounding seawater to secrete a sizeable calcium carbonate skeleton which is external to the polyps themselves. Depending on the species, polyps are positioned within individual cups in the skeleton, or in rows within long grooves in the skeleton. The polyps can retract into the skeleton for protection.

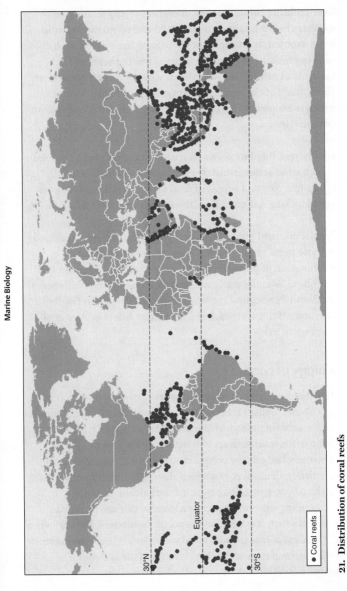

21. Distribution of coral reefs

- Coral reefs

30°N

Equator

30°S

One of the remarkable attributes of reef-building corals is that, although they are animals, functionally they perform in many ways like plants, which explains why they only flourish in well-lit environments. This is because all reef-building corals have entered into an intimate relationship with plant cells. The tissues lining the inside of the tentacles and stomach cavity of the polyps are packed with photosynthetic cells called zooxanthellae, which are photosynthetic dinoflagellates, a group of unicellular organisms generally found free-living in seawater (see Figure 22). One square centimetre of coral tissue can contain several million zooxanthellae cells.

Reef-building corals 'cultivate' the zooxanthellae for food. They do not directly consume the zooxanthellae—instead they chemically control the density of the zooxanthellae in their tissues and

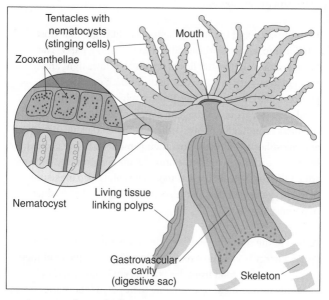

22. **Anatomy of a coral polyp**

stimulate the zooxanthellae to secrete some of the organic compounds that they synthesize through photosynthesis directly into their gut tissues. Depending on the species, corals receive anything from about 50 per cent to 95 per cent of their food from their zooxanthellae.

The corals acquire their zooxanthellae in several ways. When the polyps bud asexually, each new polyp retains some zooxanthellae. Coral polyps can also reproduce sexually and, in this case, the polyp incorporates some of its zooxanthellae into each egg it produces. Some young corals are not endowed with zooxanthellae and must acquire them from the external environment as they grow. In this case the coral secretes a chemical into the seawater that attracts the required dinoflagellates, which are ingested and incorporated into the coral's own cells. The coral then surrounds each dinoflagellate algal cell with a special membrane and begins to control its metabolism.

Both the coral and the zooxanthellae benefit from their close relationship, although the coral is the controlling party. The beauty of the relationship lies in the way it allows tight and very efficient nutrient recycling between both parties. The photosynthetic zooxanthellae, protectively sheltered within the coral's tissues, acquire a constant supply of waste metabolic products essential for photosynthesis—carbon dioxide, nitrogen, and phosphorus—directly from their coral host. In the presence of light they turn these nutrients into organic compounds, some of which the coral 'steals' back as food. The corals also make use of the oxygen generated by the zooxanthellae as a by-product of photosynthesis.

Although the zooxanthellae provide a large proportion of the coral's energy needs, most reef-building corals supplement their diet by capturing food from the external environment. Corals generally feed at night by extending their polyps above the skeleton, which is why coral colonies appear more 'furry' at night.

The mouth of each polyp is surrounded by a ring of tentacles equipped with special 'stinging' cells, called nematocysts (see Figure 22), that eject both poisonous and sticky threads that subdue small animals, mainly zooplankton, that the coral colonies feed on. Corals also secrete threads of sticky mucus that collect fine particles, which are then drawn into the polyps' mouths.

Types of coral reefs

Corals grow very slowly—in the order of a few centimetres per year—but over long periods of time form massive structures, in fact some of the largest biologically derived structures on the planet. There are three main types of coral reef structures—atolls, fringing reefs, and barrier reefs (see Figure 23).

Atolls are common in the tropical Indian and Pacific Oceans and are associated with oceanic islands. The creation of an atoll is initiated when reef-building corals colonize the sides of a newly created volcanic island to form a 'fringing' reef. Such newly formed islands often slowly sink because of the massive weight they exert on the underlying sea floor. As they sink, the corals

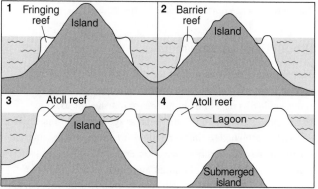

23. Stages in the formation of an atoll reef

surrounding the island continue to grow upwards on a bed of calcium carbonate that they secrete—a platform of limestone that gets deeper and deeper over time. These reefs eventually become separated from the island by a channel of seawater and are known as 'barrier' reefs. As the island itself eventually disappears beneath the surface of the ocean, the coral reef continues to grow upwards from its island base towards the ocean surface, creating a ring- or semi-ring-shaped structure around a seawater-filled lagoon—the atoll reef.

The living crown of an atoll can rest on an enormously thick layer of coral-created limestone. Bore holes drilled at Eniwetok Atoll in the Marshall Islands penetrated through close to 1,300 metres of limestone before hitting volcanic rock marking the top of the volcanic island on which the atoll originated. It would have taken about sixty million years of reef growth to create a limestone cap this thick.

Fringing and barrier reefs can also develop alongside continental land masses. Fringing reefs are separated from the coast by a narrow channel, whereas barrier reefs occur at a larger distance off the coast. Barrier reefs can be very large structures, the largest being the Great Barrier Reef which stretches for about 2,600 kilometres off the northeast coast of Australia.

Productivity of coral reefs

The corals are the backbone of the reef ecosystem, creating a complex, three-dimensional habitat that supports a truly remarkable diversity and abundance of marine life. Marine invertebrates are prolific. Some, such as sponges, sea fans, and soft corals, live attached to the reef. Others are more mobile, such as sea urchins, sea cucumbers, sea stars, crabs, shrimps, and sea slugs. Colourful fish are abundant and conspicuous, and on a healthy reef large predatory fish such as groupers, barracuda, and sharks are common.

Healthy coral reefs are very productive marine systems. This is in stark contrast to the nutrient-poor and unproductive tropical waters adjacent to reefs. Coral reefs are, in general, roughly one hundred times more productive than the surrounding environment, and for this reason are often referred to as oases in a tropical marine desert.

It is at first difficult to comprehend how coral reefs can be so productive since there are no obvious plants on a reef to create a base of primary productivity. However, a large mass of primary producers is present in the form of the microscopic zooxanthellae hidden within the tissues of the corals themselves. Other kinds of algae are also widespread on a coral reef, including microscopic green algae that bore into coral skeletons, red and green coralline algae that form a ubiquitous encrusting layer on exposed hard surfaces, and various forms of delicate calcareous algae that form turf-like layers on parts of the reef. All this adds up to a large, although somewhat inconspicuous, mass of photosynthetic organisms on the reef.

The waters flowing over a coral reef are well lit but very nutrient poor. So where do the nutrients come from to support these highly productive tropical marine oases? As it turns out, coral reefs are superb nutrient sinks, able to scavenge the available nitrogen and phosphorus from their nutrient-poor surroundings, and then retain and use these nutrients very efficiently. Reef algae are able to absorb some of the scarce nutrients directly from the seawater flowing over the reef. In addition, the corals obtain some nutrients from the zooplankton and dead organic particles that they filter from the seawater as a supplementary food source. Nitrate is also created by nitrogen-fixing bacteria living in association with the corals and other reef organisms, or free-living in the seawater. Once acquired, these precious nutrients are recycled very efficiently back and forth between the plants and animals on the reef, with minimal loss to the surroundings. The ultimate example of this is the tight recycling of nutrients between zooxanthellae and the tissues of their coral hosts.

Exposed fleshy algae occur sparsely on a healthy coral reef. They are readily consumed by herbivorous fish, such as damselfish, butterfly fish, and surgeonfish, which are normally very abundant on a reef. Sea urchins, such as the menacing-looking, long-spined, black sea urchin, *Diadema*, are also efficient grazers. These animals play a very important role in maintaining a healthy coral reef system by preventing the otherwise fast-growing fleshy algae from overgrowing and smothering the corals.

Corals are not completely immune to predation despite their protective external skeletons. Several types of fish, known as corallivores, are well adapted to feeding on corals. Some, such as triggerfish, filefish, and butterfly fish, are able to pluck entire coral polyps from a colony. The forceps-like mouths of butterfly fish, which are armed with numerous small teeth, are well suited for this. Others, such as surgeonfish and parrotfish, bite or rasp off and ingest pieces of coral, together with the skeleton, and digest the coral tissue, the algae in the coral skeleton, and any encrusting coralline algae associated with the coral. The beak-like mouths of parrotfish are well adapted for this sort of feeding behaviour. The remains of the coral digested by parrotfish are excreted as sand, which accumulates in pockets on the reef, and helps form the sandy beaches associated with some reef systems.

Sexual reproduction in corals

Reef-building corals are able to disperse from the parent colony and colonize new habitats through sexual reproduction. Many species of coral are hermaphroditic—able to produce both eggs and sperm on the same colony. In others species, separate male and female colonies are the norm. The majority of coral species employ what is termed broadcast spawning—they eject huge numbers of sperm and eggs into the surrounding seawater where fertilization takes place. The fertilized egg develops into a tiny ciliated larval form, called a planula, which is transported by currents for a number of days or weeks, depending on the species.

When the larvae detect favourable conditions they swim to the bottom, where they attach themselves and start a new colony.

Coral colonies of the same species in the same area tend to spawn in synchrony—a behaviour that appears to have evolved to ensure that there is a sufficient concentration of gametes in the seawater to result in reasonable rates of successful fertilization. The breeding season of coral colonies can be controlled by factors such as seasonal changes in sea temperature or day length, which serve to bring the corals into breeding condition at the same time. The actual spawning of the ripe colonies is triggered by different factors, including lunar periodicity, a decrease in light levels at sunset, or chemical cues released into the water by other colonies of the same species.

In some regions, many different species of coral spawn simultaneously in a spectacular mass spawning event in which the seawater becomes literally saturated with coral gametes, which form distinct slicks on the ocean surface. On parts of the Great Barrier Reef, millions of colonies of dozens of species of coral spawn together over the course of a few nights during the southern hemisphere spring or early summer after a full moon. In this case, these 'primitive' corals have the rather extraordinary ability to sense moonlight and work out the phase of the moon.

Physical and biological disturbance

Though massive in structure, coral reefs are not immune to periodic large-scale physical disturbances. The large waves created by tropical hurricanes and typhoons passing in the vicinity of a reef routinely break up and overturn extensive areas of living coral. Large tsunami waves generated by major earthquakes, such as the 26 December 2004 earthquake near Indonesia, also cause large-scale damage. Coastal reefs are also periodically inundated by freshwater run-off following flood events, which can kill large numbers of corals. The rate of recovery from damage of this kind is variable, but generally of the order of decades or longer.

Coral reefs are also susceptible to large-scale biological disturbance. Population explosions of the 'crown-of-thorns' sea star, *Acanthaster planci*, routinely result in the destruction of reefs in the Pacific and Indian Oceans and in the Red Sea. The crown-of-thorns is a very large sea star, reaching a diameter of half a metre, and is a specialist feeder on coral polyps. It normally occurs at very low densities on a reef. Each animal on its own can strip the tissue from a square metre or so of living coral a month, which is generally not enough to harm the reef. However, when densities exceed about thirty sea stars per hectare of reef, the sea stars are able to consume coral at a rate faster than it can grow, and the results are disastrous for the reef. During the course of a crown-of-thorns infestation, which can last for many years, with the sea stars roving from one reef to the next, large areas of coral are denuded, leaving behind bare skeletons (see Figure 24). Recovery following such an outbreak is a slow process—typically taking several decades—and reefs can be re-infested with crown-of-thorns before they are completely recovered.

Crown-of-thorns outbreaks were first documented in the 1950s and have been observed regularly since then in many different places. Such outbreaks likely occurred naturally in the past, but the frequency and size of outbreaks are now increasing, which suggests a human influence is now involved. The crown-of-thorns' array of venomous spines makes it an unpalatable meal for most potential predators. However, several species of fish are able to eat it, as well as a large marine snail, the triton. It has been proposed that overfishing of its fish predators, together with overzealous collection of the triton, whose shell is prized by souvenir hunters and shell collectors, has allowed crown-of-thorns populations to explode in many areas. It has also been suggested that such population explosions are linked to run-off from land following abnormally heavy rainfall, which flushes excessive nutrients from agricultural land into coastal waters. The nutrients stimulate blooms of phytoplankton which create a plentiful source of food for the young planktonic larval stage of the crown-of-thorns. This

24. Crown-of-thorns sea star feeding on coral colonies. The corals to the right of the picture have been stripped of their polyps by the sea star, leaving behind bare skeleton

results in unusually high rates of survival of the young stages leading to a pulse of high juvenile recruitment and a population explosion some years later.

Though the causes of crown-of-thorns outbreaks are yet to be fully understood, it is clear that infestations are causing considerable damage to coral reefs in many areas. The Great Barrier Reef has been subject to three successive waves of crown-of-thorns outbreaks since the early 1960s. The latest outbreak started in 1993 and persists to this time, with about 15 per cent of the reef affected and showing dramatic reductions in coral cover. Once the sea stars have moved on from a specific area, the corals of the Great Barrier Reef have so far been able to recover to a state similar to their pre-outbreak condition, generally within 10–15 years.

Corals can also be affected by a range of diseases which can cause discoloration of the coral tissues, tumours, and tissue death. Little is known about the causes and effects of these diseases, although they can be associated with infections by various bacteria, fungi, algae, and worms.

Human impacts

Although subject to a range of natural physical and biological disturbances, reef-building corals have persisted for many millions of years. Now, however, coral reefs are under serious threat from a wide range of human disturbances. Overfishing constitutes a significant threat to coral reefs at this time. About an eighth of the world's population—roughly 875 million people—live within 100 kilometres of a coral reef. Most of these people live in developing countries and island nations and depend greatly on fish obtained from local coral reefs as a food source. It is not surprising, then, that unsustainable fishing is a rampant problem on many coral reef systems around the planet.

The larger, high-value, predatory fish, such as groupers, snappers, trevally, and humphead wrasse, are the first to be targeted by fishers on a healthy coral reef. These rapidly become depleted and fishers then start to fish down the food chain out of necessity and target mainly herbivorous fish. The end result is a coral reef inhabited by small fish difficult to catch and of little food value, and virtually devoid of the large predatory species such as sharks and groupers that would have once been common, as well as the larger herbivorous fish.

Coral reefs in this state become less resilient and much more vulnerable to other perturbations in the system. The coral reefs of the Caribbean region provide a classic example of the destabilizing effects of overfishing. Most of the Caribbean island nations are highly populated and the adjacent reefs have been subject to heavy fishing pressure for many decades—at least 60 per cent of the region's coral reefs are severely overfished at this time and large predatory and herbivorous fish are rare.

At first, the removal of herbivorous fish from the system was compensated by an increase in the numbers of *Diadema* sea urchins, which continued to graze back and control the amount of fleshy algae on the reef. However, beginning in 1983, a sea urchin disease spread rapidly through the Caribbean and killed almost all *Diadema* throughout the region within a couple of years. Now that virtually all herbivores were gone, fleshy algae flourished and quickly outcompeted the corals.

Over the course of a decade, the coral reefs throughout much of the Caribbean were transformed from luxurious coral-dominated systems to drab seaweed-dominated systems lacking the colour, diversity, and complexity of a healthy coral reef. Once fleshy algae become established on a coral reef, the regrowth of corals is severely disrupted. Thus, unfortunately, this massive shift appears to be for all intents and purposes permanent in the face of continued fishing pressure and other stresses to the coral reef systems of the Caribbean. Coral reefs around the world are subject to similar overfishing pressures and similar transformations as took place in the Caribbean are now widespread.

Apart from overfishing, humans impact locally on coral reefs in many other ways. Some of the fishing practices are very harmful. Once the large fish are removed from a coral reef, it becomes increasingly more difficult to make a living harvesting the more elusive and lower-value smaller fish that remain. Fishers thus resort to more destructive techniques such as dynamiting parts of the reef and scooping up the dead and stunned fish that float to the surface. People capturing fish for the tropical aquarium trade will often poison parts of the reef with sodium cyanide which paralyses the fish, making them easier to catch. An unfortunate side effect of this practice is that the poison kills corals.

Corals require clean, clear seawater to thrive and live in environments where the nutrient concentration is naturally very low. They are thus particularly susceptible to any degradation of water quality arising from coastal developments and land-use change. Sediment run-off from agricultural land, deforested areas, and from earthworks during coastal construction activities decreases water clarity and coats corals with sediment, reducing the amount of light the corals can receive, and also smothering the polyps. Even very small increases in nutrient concentrations can stress corals by stimulating elevated levels of phytoplankton, which reduce water clarity and light penetration. Increased nutrients also encourage the growth of coral-smothering seaweeds. Untreated sewage discharge is thus an obvious threat; so too are nutrients in agricultural run-off.

There are now concerns that run-off from agricultural land is affecting the resilience of parts of the Great Barrier Reef. Freshwater, sediment, and nutrients drain into the Great Barrier Reef system from a huge catchment area of about 424,000 square kilometres. Cattle grazing takes place throughout much of this catchment and sugar cane is grown on other parts of the catchment, particularly adjacent to waterways on fertile coastal floodplains. Rivers now carry roughly fourteen million tonnes of sediment, 49,000 tonnes of nitrogen, and 9,000 tonnes of phosphorus into the Great Barrier Reef region each year. In addition, herbicides used in sugar cane growing are present in the run-off. It has been estimated that sediment run-off has increased three- to tenfold, nitrogen run-off at least twofold, and phosphorus run-off at least threefold in comparison to when the catchment was undeveloped prior to 1850.

It is difficult to assess the overall effects of this run-off on the Great Barrier Reef because there is no information on what the reef was like prior to the development of the catchment—the Reef has only been seriously monitored for less than thirty years. There is concern, however, that reefs within about 10 kilometres of the coast are now at risk from nutrient enrichment and that reefs further offshore are being impacted in ways that will ultimately reduce their resilience to other impacts. In response, efforts are under way to develop and evaluate new land management practices that will reduce the pollutants in run-off, such as more efficient use of fertilizers, limiting the use of herbicides, and re-establishing riparian vegetation along the edges of rivers and streams to help filter out sediments and nutrients.

Coral reefs have only been seriously studied since the 1970s, which in most cases was well after human impacts had commenced. This makes it difficult to define what might actually constitute a 'natural' and healthy coral reef system, as would have existed prior to extensive human impacts. Coral reef biologists have attempted to obtain a clearer picture of what an un-impacted coral reef system is like, and in so doing reset our biased 'baseline', by studying reefs on uninhabited atolls in the remote Line Islands, which are situated in the central Pacific Ocean 1,600 kilometres south of Hawaii. They then compared what they found on the uninhabited atolls with increasingly populated atolls in the same chain of islands which have been subjected to various levels of fishing pressure and pollution.

They found that the coral reefs on the uninhabited atolls are dominated by large numbers of apex predators—large fish such as sharks, jacks, red snappers, and groupers; live corals cover nearly 100 per cent of the bottom and fleshy algae are virtually absent. On these reefs an extraordinary 85 per cent of the fish biomass consists of large predators, about three-quarters of which are sharks. This highly 'top heavy' biomass of predatory fish is sustained by a rapid turnover of quickly reproducing and growing prey fish such as butterfly fish, parrotfish, and damselfish. Coral reefs on the populated atolls are quite different. Here predatory fish are rare and the reefs are dominated by large numbers of small, aquarium-sized herbivorous fish—the 'bottom heavy' pattern of fish biomass that we have come to think of as characteristic of coral reefs. These reefs have much lower levels of coral cover and much larger amounts of fleshy algae. Thus, these few remnant coral reef systems on remote uninhabited atolls provide us with a glimpse of what most coral reefs probably looked like hundreds of years ago prior to widespread human influence.

The human threats to coral reefs discussed thus far are local or regional in scale. The ultimate peril to coral reefs is planet-wide in scale—ocean warming and acidification resulting from human-induced climate change.

Corals are very sensitive to sea temperature, and small increases above normal summer maximum temperatures result in stress. Temperature-stressed corals undergo 'bleaching', in which they expel the zooxanthellae from their tissues. Without the zooxanthellae, the coral tissues become transparent and reveal the white limestone skeleton beneath. If temperature stress is of low level and short duration, corals can reacquire their zooxanthellae and survive, although they may be more susceptible to other stresses, such as disease. Highly temperature-stressed corals are not able to reacquire zooxanthellae sufficiently rapidly and without them they die.

Occasional, small-scale episodes of coral bleaching are a natural phenomenon on coral reefs. However, coral bleaching events have been occurring with much greater frequency and intensity in the last decade compared to the previous decade as extreme sea temperatures occur more frequently. A major mass bleaching event in 1998 killed an estimated 16 per cent of all corals on the planet, with 60–90 per cent of all corals dying in the central and western Indian Ocean. About three-quarters of these affected reefs were able to recover to some extent. Modelling of future sea surface temperatures, combined with knowledge of coral physiology, suggests that by the 2050s severe coral bleaching events will be occurring just about every year in almost all coral reef systems on the planet.

The increasing concentrations of carbon dioxide in the atmosphere are not only causing ocean warming, but are also making seawater more acidic. In the near term, ocean acidification makes it more difficult for corals to manufacture their calcium carbonate skeletons. To make their skeletons, corals need to combine calcium ions and carbonate ions dissolved in the surrounding seawater. However, acids react with carbonate ions, converting them into bicarbonate ions, making them less available to the coral. The corals thus have to expend more energy to produce their skeletons, which slows their growth and causes stress, making them more susceptible to other stresses such as temperature and disease. Continued acidification will eventually cause corals, and other marine organisms with calcium carbonate structures, to stop growing altogether and, at some point, cause their skeletons to slowly dissolve. By 2030, atmospheric carbon dioxide levels will reach at least 450 parts per million (ppm). At this level it is predicted that coral growth will be severely reduced. By 2050, carbon dioxide levels will be around 500 ppm. At this level, only a few areas of the ocean will be suitable for reef-building corals to survive.

In the face of such a variety of human-related regional and planet-wide impacts, the overall health of coral reef systems is deteriorating rapidly. The latest assessment is that 19 per cent of the planet's natural coverage of coral reefs has been permanently lost; another 15 per cent is likely to disappear within the next ten to twenty years; and another 20 per cent is under threat of loss in twenty to forty years. Thus, in the not too distant future, more than half of the planet's coral reefs appear destined to disappear.

And what is even more disturbing is that these estimates only take account of local threats, and do not consider the threat posed by elevated sea temperatures resulting from human-induced climate change. When this is included, about 75 per cent of the planet's reefs are threatened. Even the most remote and regionally unimpacted coral reefs will not escape the effects of a warming ocean and ocean acidification, although their healthy condition should make them more resilient to these planet-scale impacts. Given the variety, scale, and synergistic effects of human-related stresses on coral reefs, the sad reality is that in most places on the planet coral reefs as we now know them are destined to disappear by the 2050s or so.

Mangroves

Mangrove is a collective term applied to a diverse group of trees and shrubs that colonize protected muddy intertidal areas in tropical and subtropical regions, creating mangrove forests, or mangals.

Two of the most common species of mangrove trees are the red mangrove, *Rhizopora mangle*, and the black mangrove, *Avicennia germinans*. The red mangrove can be distinguished by its characteristic tangle of prop roots that help support the tree in the soft substratum. The black mangrove has a more typical tree-like trunk surrounded by a mass of peg-like structures, called pneumatophores, which arise from a system of roots buried in the sediment and radiating out from the trunk.

Mangroves inhabit a very harsh environment consisting of seawater-logged sediments which are anaerobic, or lacking in oxygen. To deal with the lack of oxygen, mangrove trees have evolved root systems that are adept at extracting oxygen from the air or from the seawater when they are submerged. The prop roots of the red mangrove are covered with small nodular structures called lenticels through which oxygen is supplied to the underground root system. The peg-like pneumatophores of the black mangrove, which are also covered with lenticels, act like snorkels, drawing oxygen from the air or surrounding seawater (see Figure 25).

Mangrove trees have also adapted to growing in salty sediments. They deal with excess salt in a number of ways. Mangrove tree roots and stems have special tissues which act as a barrier to reduce the amount of salt that enters the plant. Nonetheless, some salt does penetrate the plant, which can tolerate high

Marine life in the tropics

25. Black mangrove tree, *Avicennia germinans*, St Thomas, US Virgin Islands, showing an extensive system of pneumatophores

concentrations in its sap—salt concentrations in the sap of mangroves can be ten to a hundred times greater than is found in normal plants. In addition, the leaves of the black mangrove possess special glands that concentrate and excrete excess salt—the salt crystals collect on the under surface of the leaves and are washed off during rain. Mangroves also concentrate salt in old leaves, bark, flowers, and fruit, which take away the salt when they drop off the tree.

Mangrove trees produce flowers pollinated by the wind or by bees. The flowers produce seeds which germinate into seedlings which grow into young trees with distinctive cigar-shaped stems while still on the tree. These young trees, or propagules, are ready to produce roots as soon as they fall from the parent tree and encounter a suitable habitat. The propagules are buoyant and can survive in seawater and can drift with the currents for more than a year. Once stranded on a suitable shore they quickly take root and grow.

Mangroves form a complex and productive habitat. Few organisms, except for a few crabs, can graze directly on the mangrove leaves. However, mangroves are constantly dropping off dead leaves and branches, which are broken down by bacteria and fungi, and form the basis of a productive food web. Crabs and shrimps and other organisms graze on this detrital material and are then eaten by fish, turtles, and shorebirds.

Mangroves are of great importance from a human perspective. The sheltered waters of a mangrove forest provide important nursery areas for juvenile fish, crabs, and shrimp. Many commercial fisheries depend on the existence of healthy mangrove forests, including blue crab, shrimp, spiny lobster, and mullet fisheries. Mangrove forests also stabilize the foreshore and protect the adjacent land from erosion, particularly from the effects of large storms and tsunamis. They also act as biological filters by removing excess nutrients and trapping sediment from land run-off before it enters the coastal environment, thereby protecting other habitats such as seagrass meadows and coral reefs.

Large-scale destruction of mangroves occurs naturally as a result of hurricanes and typhoons, which uproot trees, or smother the roots with excess sediment. Mangroves can generally recover from such events in two or three decades. However, most of the destruction of mangroves is now brought about by human activity. Mangroves are heavily harvested by humans for timber, firewood, and for the production of charcoal. They are also routinely drained to make way for coastal development. Mangrove forests are also frequently converted into large ponds for the culture of shrimp and fish or for the production of salt.

As a result of this heavy human pressure, mangrove forests are disappearing rapidly. In a twenty-year period between 1980 and 2000 the area of mangrove forest globally declined from around 20 million hectares to below 15 million hectares. In some specific regions the rate of mangrove loss is truly alarming. For example, Puerto Rico lost about 89 per cent of its mangrove forests between 1930 and 1985, while the southern part of India lost about 96 per cent of its mangroves between 1911 and 1989. Concerted conservation and management efforts will be required if there is to be any chance of lessening the pace of mangrove destruction in the face of a rapidly growing human population. As much as possible, existing stands of healthy mangroves will require outright protection. Additionally, policies and strategies will need to be developed and implemented to begin to regenerate degraded mangrove habitats in critical areas.

Chapter 6
Deep-ocean biology

The deep ocean is by far the largest ecosystem on the planet. To put the vastness of this region into perspective, consider that about 80 per cent of the entire volume of the Global Ocean, or roughly one billion cubic kilometres, consists of seawater with depths greater than 1,000 metres and this region encompasses 79 per cent of the planet's habitable living space. Yet, because of its inaccessibility, this vast region is the least investigated and understood environment on the planet.

The physical environment of the deep ocean

The deep ocean is a permanently dark environment devoid of sunlight, the last remnants of which cannot penetrate much beyond 200 metres in most parts of the Global Ocean, and no further than 800 metres or so in even the clearest oceanic waters. The only light present in the deep ocean is of biological origin—weak and sporadic flashes of bioluminescence created by a chemical reaction in specialized organs present in the bodies of some deep-ocean animals.

Extreme pressure is another defining characteristic of the deep ocean. Seawater is a heavy substance and a column of seawater 10 kilometres high—typical of the deeper parts of the Global

Ocean—exerts a pressure of 10,000 tonnes per square metre, which is about the weight of 55 jumbo jets.

Except in a few very isolated places, the deep ocean is a permanently cold environment, with sea temperatures ranging from about 2°C to 4°C. Since the deep ocean is below the oxygen minimum zone, usually present at depths from about 200 to 1,000 metres, dissolved oxygen concentrations are generally more than adequate to support life.

Food is scarce in the deep ocean. Since there is no sunlight, there is no plant life, and thus no primary production of organic matter by photosynthesis. The base of the food chain in the deep ocean consists mostly of a 'rain' of small particles of organic material sinking down through the water column from the sunlit surface waters of the ocean. This reasonably constant rain of organic material is supplemented by the bodies of large fish and marine mammals that sink more rapidly to the bottom following death, and which provide sporadic feasts for deep-ocean bottom dwellers.

The picture of the deep ocean that thus emerges is one of a cold, dark, extremely pressurized, food-limited environment—in short, a very harsh and extreme environment from a human perspective. And yet, this is the largest and most typical ecosystem on the entire planet, and one which possesses a great diversity of marine life specialized for living under such conditions.

Adaptations of deep-ocean animals

Deep-ocean animals have adapted to the lack of light in a number of ways. For fish living in the sunlit photic zone, their eyes are useful for navigation, finding food and mates, and avoiding predators. In the pitch black of the deep ocean, eyes become a more problematic asset. Some deep-ocean fish have greatly overdeveloped eyes in comparison to fish living in the photic zone.

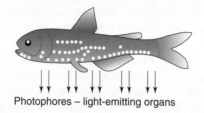

Photophores – light-emitting organs

26a. Lanternfish showing photophores along the side and belly

Such fish often come equipped with an array of light-producing organs, or photophores, as well. A good example of this is the lanternfish, which is common in the upper regions of the deep ocean. Lanternfish possess rows of photophores along their belly and sides which emit weak blue, green, or yellow light (see Figure 26a). These are arranged in a pattern specific to each species of lanternfish and, in some species, the pattern differs between males and females. Their large eyes probably evolved to allow these fish to detect the low-intensity light being emitted from other members of the same species. This would be useful for identifying a mate in the vastness of the deep ocean. On the other hand, some deep-ocean fish have small, degenerate eyes or have lost their eyes entirely. These fish are likely equipped with a powerful sense of smell and release a chemical into the sea that attracts potential mates.

The anglerfish has solved the problem of finding a mate in the vastness of the deep ocean through an unusual adaptation—the male of the species has been reduced to a tiny parasitic form that attaches itself permanently to the much larger female (see Figure 26b). His mouth fuses to her body and his blood vessels merge with hers. Thus, a male is always present to fertilize the eggs of the female, eliminating the need for the female to find a male at breeding time. Of course, the dwarf male will have to locate a female in the first instance, presumably by smell, but once attached, the problem of finding another mate is solved.

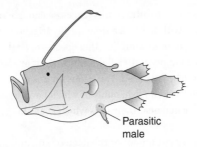

26b. Anglerfish with attached parasite

26c. Black swallower with an engulfed prey fish

Since food is a scarce commodity for deep-ocean fish, full advantage must be taken of every meal encountered. This has resulted in a number of interesting adaptations. Compared to fish in the shallow ocean, many deep-ocean fish have very large mouths capable of opening very wide, and often equipped with numerous long, sharp, inward-pointing teeth. Good examples include the gulper eel, anglerfish, loosejaw, and black swallower (see Figure 26c). These fish can capture and swallow whole prey larger than themselves so as not to pass up a rare meal simply because of its size. These fish also have greatly extensible stomachs to accommodate such meals.

Some deep-ocean fish have evolved ways to lure scarce prey. Anglerfish possess a long, flexible tentacle—a kind of fishing pole—that extends upwards between the eyes, at the end of which is a fleshy growth called the 'esca' (see Figure 26b). In some species, the esca can be wiggled so as to resemble a small fish which acts as a lure to attract prey fish close enough to be engulfed whole. Contact with the esca automatically triggers the jaws to respond in trapdoor fashion. In other species of anglerfish the tip of the esca is equipped with a light-emitting organ that lures prey close enough to be devoured. Variations on this theme are common in deep-ocean fish, many of which have light-emitting appendages located near their jaws which act as lures.

In the pelagic environment of the deep ocean, animals must be able to keep themselves within an appropriate depth range without using up energy in their food-poor habitat. This is often achieved by reducing the overall density of the animal to that of seawater so that it is neutrally buoyant. Thus the tissues and bones of deep-sea fish are often rather soft and watery. The deep-ocean pelagic environment is also dominated by gelatinous animals such as jellyfish, siphonophores, ctenophores, and salps whose body density is close to that of seawater.

Not surprisingly, deep-ocean animals have evolved adaptations for life in a highly pressurized environment. These high pressures not only impact on deep-ocean animals structurally, but also have profound effects on their physiology and biochemistry. High pressure affects the physiology of cell membranes by compressing them, expelling fluid, and making them more rigid, and thus less capable of channelling nutrients and wastes in and out of the cell. There is evidence that deep-ocean organisms have developed biochemical adaptations to maintain the functionality of their cell membranes under pressure, including adjusting the kinds of lipid molecules present in membranes to retain membrane fluidity under high pressure. High pressures also affect protein molecules, often preventing them from folding up into the correct shapes for them to function as efficient metabolic enzymes. There is evidence that deep-ocean animals have evolved pressure-resistant variants of common enzymes that mitigate this problem.

Animals of the deep-ocean seabed

Almost all of the food that supports the benthic animals of the deep ocean is derived from primary production in the photic zone. This food is in the form of dead organic material that sinks from the ocean surface as 'marine snow'. This 'snow' consists of small sticky clumps, or aggregates, of organic particles that include phytoplankton cells, dead zooplankton, and the faecal pellets produced by zooplankton. These aggregates sink slowly through the water column at a rate of about 100 to 200 metres per day. It thus takes several weeks for the aggregates to reach the deep ocean. Along the way, some of the nutritional value present in the aggregates is extracted by bacteria in the water column; thus, the deeper an aggregate sinks, the more it becomes depleted of its nutritious substances. When this organic material finally reaches the ocean bottom, some of it is available to be consumed by benthic animals equipped to filter organic particles suspended in the layer of seawater just above the sea bed. The remainder is deposited into the sea bed itself where it accumulates and can be consumed by deposit-feeding animals living on or in the sediments, as well as by sedimentary bacteria. These benthic particle feeders can in turn be eaten by predatory invertebrates, and the benthic fauna in general can be preyed upon by larger animals such as bottom-dwelling demersal fish. Because organic material accumulates in the bottom over very long periods of time, the deep-ocean sediments contain more organic material compared to the overlying seawater and can support a comparatively larger biomass of animals.

As a result of its remoteness, only about 5 per cent of the deep-ocean bottom has been explored with remotely operated robotic vehicles, and less than about 0.01 per cent has been sampled in detail. The deep-ocean bottom was once considered to be a biological desert, but as more samples were obtained, and more observations made from submersibles and through the cameras of remotely operated vehicles, it became clear that a spectacular diversity of animals inhabit the deep-ocean sea floor. These include many different species of suspension-feeding invertebrates such as sponges, sea lilies, sea pens, sea anemones, sea fans, and fan worms; deposit-feeding animals such as worms, sea cucumbers, brittle stars, sea urchins, and clams; and predators such as sea urchins, sea stars, sea anemones, amphipods, and octopuses (see Figure 27).

Marine biologists have struggled for some time to explain the high diversity of the deep-ocean benthic community. For the most part, the bottom of the deep ocean is a flat, seemingly featureless, sedimentary plain lacking the obvious habitat complexity of a coral reef or kelp bed, celebrated for their diversity of marine life.

The pattern of species diversity of the deep-ocean benthos appears to differ from that of other marine communities, which are typically dominated by a small number of abundant and highly visible species which overshadow the presence of a large number of rarer and less obvious species which are also present. In the deep-ocean benthic community, in contrast, no one group of

27. Sea cucumber on the deep-ocean sea floor

species tends to dominate, and the community consists of a high number of different species all occurring in low abundance.

This particular pattern of high species diversity appears to result from a number of factors peculiar to the deep-ocean benthic system. One is related to food. Since food is scarce on the deep-ocean floor and consists of the same type of food—organic material derived from the surface—it has been hypothesized that deep-ocean benthic animals are all eating pretty much the same food type and thus competing heavily for that food source. As a result, no one species or group of species is able to grow its population to the point where it can outcompete other species, and so many species are able to coexist.

In contrast, there is some evidence that deep-ocean deposit feeders are actually able to selectively feed on organic particles of a certain size range. Thus some are specialized to ingest tiny food particles, others medium-sized ones, and yet others larger ones. In this way a diversity of species could coexist in the same spot and species diversity would be higher in those areas where there is a large size range of organic particles present in the sediments.

Another factor put forth is the vast size of the deep-ocean habitat. In general, species diversity increases with the size of a habitat—the larger the area of a habitat, the more species that have developed ways to successfully live in that habitat. Since the deep-ocean bottom is the largest single habitat on the planet, it follows that species diversity would be expected to be high. Furthermore, the deep ocean is a comparatively stable and potentially ancient system where species are not as likely to become extinct because of physical perturbations and there has been plenty of time for new species to evolve and accumulate.

It has also been noted that the deep-ocean bottom is not as monotonous as once thought and is punctuated by physical features such as mounds, holes, and tracks whose origin is unclear, but likely the result of some form of biological disturbance. These features persist for very long periods because of the stability of the deep-ocean environment. Such features provide a measure of microhabitat variability that probably also promotes species diversity.

In summary, it is fair to say that the factors responsible for creating the high diversity of the deep-ocean benthos are complex and will not be better understood until much more sampling and detailed study of this habitat take place.

Whale falls and other deep-ocean food riches

Although the deep ocean is one of the most food-limited environments on the planet, large packets of food do occasionally and unpredictably arrive on the bottom. The bodies of large marine animals that sink rapidly to the bottom of the deep ocean in an intact condition are one source of such food bonanzas. 'Whale falls', the carcasses of dead whales on the seabed (see Figure 28), have been observed directly from deep-sea submersibles and indirectly using side-scan sonar devices. It was generally assumed that whale falls were rare events in the deep ocean but there is growing evidence that such feasts occur more frequently than assumed. It has been estimated that at any one time there are hundreds of thousands of whale carcasses on the bottom of the Global Ocean in various stages of decomposition. The distance between whale falls on the sea bed has been calculated to be in the order of 12 or so kilometres on average and perhaps much closer than this along the migration routes of more abundant whale species.

Whale falls provide a nutritional oasis for a host of deep-ocean sea animals. When these large 30–160-tonne corpses first arrive on the sea floor they are quickly located by large aggregations of mobile scavengers such as hagfish, rat-tails, and sleeper sharks which relentlessly remove much of the fleshy material over the

28. Remains of a whale fall in Monterey Canyon off the coast of California. This photo was obtained during a February 2002 dive using a remotely operated vehicle, the *Tiburon*. The 'fuzz' represents thousands of deep-sea worms growing on the whale bones

course of months to years, depending on the size of the whale. The remains of the carcass are then colonized by worms and crustaceans that over the next months or years consume the fats present in the bones of the whale. These include specialized worms, called boneworms, which extend root-like structures into the whale bones. These 'roots' contain bacteria that help digest the fats and transfer the nutrients to the worms. In the final stage, bacteria consume the remaining fats present in the bones and in the sediments beneath the carcass, a stage which lasts for decades.

Even wood is a food resource in the deep ocean. 'Wood falls' are a common occurrence and comprise trees and tree parts that have been swept into the ocean and sunk to the deep-ocean seabed. These wood falls form the basis of a specialized community where the wood is colonized by bacteria, boring clams, and wood-eating crabs. The guts of the crabs contain bacteria and fungi that assist the crab in digesting the cellulose in the wood.

Seamounts

Seamounts represent a special kind of biological hotspot in the deep ocean. Seamounts rise abruptly from the deep-ocean floor and their peaks can be thousands of metres beneath the ocean surface. In contrast to the surrounding flat, soft-bottomed abyssal plains, seamounts provide a complex rocky platform that supports an abundance of organisms that are distinct from the surrounding deep-ocean benthos.

A prolific group of suspension-feeding animals dominates the summit and flanks of seamounts, creating a dense thicket-like community comprising cold-water stony corals, sea fans, black corals, and sponges. These coral-dominated thickets create habitat for a host of other animals such as feather stars, sea lilies, brittle stars, sea stars, sea cucumbers, barnacles, sea squirts, worms, shrimp, and dense aggregations of fish. Being isolated, island-like habitats, seamounts can have very different groupings of corals and associated animals, and even nearby seamounts separated by only tens of kilometres can have quite distinct communities.

Unlike shallow-water, reef-building corals of the tropics, which obtain much of their food from their photosynthetic zooxanthellae in the presence of sunlight, deep-ocean, cold-water stony corals rely solely on filtering zooplankton and suspended organic particles from the seawater. These stony corals grow very slowly and can be several hundred years old. Other types of corals associated with seamounts show extreme longevity, with life spans of thousands of years. For example, a type of black coral, *Leiopathes*, sampled from a seamount in the Pacific Ocean, was shown using radiocarbon dating to be about 4,200 years of age, making it one of the world's longest-living animals.

The high productivity of seamount communities is driven by a number of factors which vary with the depth of the seamount and the patterns of ocean circulation in the vicinity of the seamount. Some seamounts act as a trap for large numbers of migrating zooplankton which become concentrated around the summit and provide a source of food for the abundant suspension feeders and plankton-eating fish that are present. This phenomenon is associated with a rather extraordinary and puzzling behaviour exhibited by many kinds of open-ocean zooplankton—a daily vertical migration that takes them from shallow water to depths of many hundreds of metres. During the night, zooplankton, such as krill and shrimp, stay closer to the surface, potentially to feed on phytoplankton while avoiding visual fish predators which are more abundant in shallower water. At daybreak, however, they begin a migration into deeper, colder water, perhaps to escape plankton-eating fish. On their journey, these migrating zooplankton are accompanied by squid and certain species of fish, such as lanternfish, which are very abundant in the deep ocean. At sunset, this aggregation of pelagic animals makes its way back up the water column into shallower water once again. Whatever the reason behind this energy-consuming migration, this layer of migrating animals is so dense that it shows up on sonar systems as a distinct mid-water echo called the Deep Scattering Layer (DSL). When a seamount is present at the right depth, it can get in the way of the downward-migrating DSL, trapping and concentrating the animals each night.

Seamounts can also interact with and modify the horizontal currents flowing past them. This can result in the creation of so-called 'Taylor columns', or vortices of rotating seawater that are retained over the summit of the seamount. These can further serve to retain and concentrate zooplankton around the summit and flanks of the seamount.

Seamounts support a great diversity of fish species—the latest census reveals close to 800 species have been recorded living around seamounts. In the 1960s, deep-sea trawling vessels looking for new stocks of fish began to trawl seamounts and discovered large stocks of commercially important fish species. This triggered the creation of new deep-ocean fisheries focused on seamounts. Heavily built bottom trawls are towed from the summit down the flanks of seamounts to capture the fish. Commercial fish species that are targeted include orange roughy, oreos, alfonsinos, grenadiers, and toothfish. These fish are not generally permanent residents of seamounts, but aggregate at seamounts at certain times of the year to spawn, to feed on squid and small fish, or simply to rest. They are very slow-growing and long-lived and mature at a late age, and thus have a low reproductive potential. A good example of this is the orange roughy, which is known to live for more than a hundred years and reaches maturity at around thirty years of age, with the females producing relatively small numbers of eggs. Such a life history is typical of many deep-ocean fish species.

Seamount fisheries have often been described as mining operations rather than sustainable fisheries. They typically collapse within a few years of the start of fishing and the trawlers then move on to other unexploited seamounts to maintain the fishery. The recovery of localized fisheries will inevitably be very slow, if achievable at all, because of the low reproductive potential of these deep-ocean fish species.

The destruction of fish stocks is not the only concern associated with seamount fishing. The trawling of seamounts causes extensive damage to the fragile coral communities, with the trawls bringing up not only fish, but large numbers of stony corals, black corals, and other benthic animals associated with the corals. The intensity of trawling on seamounts can be very high, with many hundreds to thousands of trawls often carried out on the same seamount. Tens of tons of coral can be brought up in a single trawl, and in one new seamount fishery it was estimated that almost one-third of the total catch consisted of coral by-catch.

Comparisons of 'fished' and 'unfished' seamounts have clearly shown the extent of habitat damage and loss of species diversity brought about by trawl fishing, with the dense coral habitats reduced to rubble over much of the area investigated.

Not surprisingly, seamount-based fisheries have become very controversial and some large food retailers have now banned the sale of orange roughy from their shelves. An increasing number of countries have begun to close some of the seamounts present in their exclusive economic zones (EEZs) to fishing. Unfortunately, most seamounts exist in areas beyond national jurisdiction, which makes it very difficult to regulate fishing activities on them, although some efforts are under way to establish international treaties to better manage and protect seamount ecosystems. It appears that the future for seamount ecosystems lies in the balance between protecting some from fishing altogether while allowing fishing on others under some form of regulation. What sort of balance is ultimately achieved remains to be seen in the face of declining fish stocks globally versus a growing world demand for fish protein.

Hydrothermal vent communities and cold seeps

The idea that all food in the deep ocean is imported from somewhere else is not strictly true. There are some very special and remarkable spots on the deep-ocean floor where food is created in place. This is a form of primary production that is not driven by the energy in sunlight but by chemical energy. Submarine hydrothermal vents represent one example of deep-ocean life driven by chemical energy.

Hydrothermal vents were discovered from the deep-diving submersible, *Alvin*, in 1977 on an ocean ridge near the Galapagos Islands at around 2,700 metres. Since then many other hydrothermal vents have been discovered in the Pacific, Indian, Atlantic, and Arctic Oceans, mostly on ocean ridges, and it has been estimated that many thousands of vents are active at any one time on the ocean floor.

Hydrothermal vents are created when seawater seeps deep down into the seabed where it reacts with hot rock to form a superheated acidic fluid that is laden with chemicals, including hydrogen sulphide (H_2S). This fluid can be ejected at high pressure back into the ocean through fissures as a hot water geyser. Generally a number of such geysers are clustered together to form vent fields that range from pool table to tennis court size.

Some geysers eject black-coloured fluid through chimney-like structures that can be tens of metres in height (see Figure 29). The vent fluid from these 'black smokers' is darkened by tiny suspended sulphur-bearing mineral particles that precipitate out of the vent fluid as it mixes with the surrounding cold seawater. The vent fluid from black smokers can exceed 400°C when it first

29. A mass of tube worms (left of the picture) bask in warm seawater near a black smoker spewing 400°C fluid at a depth of 2,250 metres

exits the seabed. Other geysers, called 'white smokers', eject a lighter-coloured fluid at lower temperatures.

Many different kinds of microorganisms live near the vents, including bacteria and archaea. Archaeans look like bacteria, but biochemically and genetically they are very distinct from bacteria. They inhabit some of the most extreme environments on the planet, including salt ponds, hot springs, and hydrothermal vents.

Bacteria and archaea are able to break down hydrogen sulphide (H_2S) to create energy to produce organic compounds from carbon dioxide that they absorb from the seawater.

$$6\ O_2 + 6\ H_2S + 6\ CO_2 + 6\ H_2O \quad \rightarrow \quad \underset{\substack{\text{organic} \\ \text{compounds}}}{C_6H_{12}O_6} + \underset{\text{sulphuric acid}}{6\ H_2SO_4}$$

These microorganisms are playing a similar role to photosynthetic organisms in the photic zone of the oceans, but are producing organic matter by means of chemosynthesis, not photosynthesis. Ultimately, however, this form of chemosynthesis based on hydrogen sulphide is linked to sunlight because the chemosynthetic microorganisms involved require oxygen to drive the chemosynthetic process (see equation above), and photosynthesis is the source of oxygen on the planet. For this reason, this form of chemosynthesis is known as aerobic chemosynthesis.

However, some of the chemosynthetic microorganisms associated with vents can use methane (CH_4), which can also be present in vent fluids, as a food source. How this is done is not yet fully understood but it appears that two kinds of microorganisms are involved, likely living in close association. One group uses methane as a carbon source and the other uses sulphate as an energy source. In this way methane can be used in combination with sulphate to produce organic compounds. Although no oxygen is directly involved in this process, this form of anaerobic chemosynthesis is probably not totally independent of photosynthesis because both methane and sulphate are derived ultimately from organic matter buried deep below the seabed that has been degraded by high temperatures. This organic matter would have been produced in ancient times by photosynthesis.

The chemosynthetic microorganisms associated with vents can be suspended in the vent plume or form a mat on the rocky bottom adjacent to the vent. These microorganisms provide the basis for an astonishing community of large animals that are abundant around the vents. These include giant clams and mussels and various species of crabs, snails, sea anemones, worms, shrimp, amphipods, stalked barnacles, octopuses, and fish. Some of these animals, such as the clams and mussels, filter the microorganisms from the seawater, while others graze on the microbial mats. These animals are in turn eaten by predators. Most of the species found at vents were previously unknown before the vents were discovered and are found only at vent systems.

Giant tube worms, called vestimentiferans, are abundant at some vent sites in the Pacific Ocean (see Figure 29). These peculiar worms, up to three metres in length, live within a white protective tube. They have no trace of a digestive system and obtain all of their nutrition from symbiotic chemosynthetic bacteria living within their tissues. These worms possess a bright red plume that can be extended out of their tube to absorb hydrogen sulphide and oxygen from the vent waters. The chemicals are then transported within the blood system of the worm to the bacteria living within the worm's tissues, which produce food for themselves and their tube worm host. The giant clams and mussels living at the vents also possess symbiotic chemosynthetic bacteria which live in the tissues of their gills. The organic matter produced by these bacteria provides most of the food required by these clams and mussels and, although they are capable of filter-feeding, they will die without the nutrition produced by their symbiotic bacteria.

Hydrothermal vents are unstable and ephemeral features of the deep ocean. Repeated observations of known vent systems show that the rate of flow and chemical composition of the vent fluids can vary over a period of months, and dead vents surrounded by the remains of vent communities have been observed. The lifespan of a typical vent is likely in the order of tens of years. Thus the rich communities surrounding vents have a very limited lifespan. Since many vent animals can live only near vents, and the distance between vent systems can be hundreds to thousands of kilometres, it is a puzzle as to how vent animals escape a dying vent and colonize other distant vents or newly created vents. It is known that some vent animals grow very fast and can reach a large body size and sexual maturity before the vent dies. On account of their large size, they can produce a large number of planktonic larvae that slow-moving deep-water currents may disperse over large distances, allowing them to colonize other vent sites. In some species, the larvae may rise to the surface where they are dispersed more rapidly by surface currents, before sinking back to the ocean floor where they may encounter a suitable vent site to colonize.

Hydrothermal vents are not the only source of chemical-laden fluids supporting unique chemosynthetic-based communities in the deep ocean. Hydrogen sulphide and methane also ooze from the ocean bottom at some locations at temperatures similar to the surrounding seawater. These so-called 'cold seeps' are often found along continental margins over a range of depths, and have been discovered in the Gulf of Mexico, off the coast of California and Alaska, and in the Sea of Japan, the latter at a depth of 5,000 to 6,500 metres.

The communities associated with cold seeps are similar to hydrothermal vent communities, with many of the animals having a close relationship with chemosynthetic bacteria that provide them with nutrition. Clams, mussels, sponges, and crabs can be very abundant, along with dense thickets of vestimentiferan tube worms. Cold seeps appear to be more permanent sources of fluid compared to the ephemeral nature of hot water vents. Thus, in contrast to hot water vent communities, cold seep communities can possess slow-growing, long-lived species. For example, some tube worms associated with cold seeps may be up to 250 years old.

Chapter 7
Intertidal life

The intertidal region of the Global Ocean is a thin strip of shoreline lying between the high and low tide marks—it is completely submerged by seawater at the highest high tides and completely uncovered at the lowest low tides. The intertidal region is occupied almost exclusively by marine organisms that have adapted to live in a very stressful physical environment influenced by exposure to air, temperature extremes, wind, and the pounding of waves. Although this region represents just a small part of the Global Ocean, it is home to a diverse and interesting marine community that people can study and enjoy on a routine basis on account of its accessibility. It is also a place where people routinely harvest seafood, and it is prone to a wide range of human impacts, including overharvesting, oil spills, coastal development, and the footsteps of thousands of visitors.

Tides

The regular rise and fall of the tides are the dominant feature of the intertidal region. The driving force behind these tides is the gravitational pull of the moon and the sun on the fluid mass of the Global Ocean. Since the moon is much closer to the earth than the sun, it has a greater effect in creating tides than the sun.

The moon causes the oceans on the side of the earth closest to the moon to bulge out slightly. Another bulge in the oceans occurs on the opposite side of the earth because, in simple terms, the earth is also being pulled towards the moon and away from the water on this far side. The earth continues to rotate beneath these two bulges and thus, in theory, any one point on the planet will pass beneath two bulges each day, which explains why tides on a shoreline often occur twice a day, roughly twelve hours apart.

The pull of the sun modulates the influence of the moon. When the earth, moon, and sun are all roughly positioned in a straight line (each month at the full and new moon), the pull of the sun is added to that of the moon and thus the tides are highest around this time—these are the so-called 'spring' tides. When the earth, moon, and sun are at roughly right angles to each other (at the first and last quarters of the moon), the pull of the sun subtracts from that of the moon and the tides are lowest at around this time—'neap' tides.

In this way the moon and sun establish the basic rhythm and height of the tides on our planet. In reality, these are modified greatly by the continental land masses, which obviously interfere with the ocean bulges, and also by the shape of the ocean basins and the peculiarities of the local shoreline. The result is that the moon and sun set up a kind of basin-scale tidal sloshing of the oceans that is modified regionally and locally to create different tidal patterns and heights at any one shoreline location. Thus, although most coastal locations have two low and two high tides per day of about the same height (termed semidiurnal tides), some have two high and two low tides of different heights each day (mixed semidiurnal tides), and a few places have only one high tide and one low tide per day (diurnal tides).

Adaptations of intertidal organisms

The tides have a profound impact on intertidal marine organisms, which are regularly submerged by the incoming tide and then exposed for varying periods of time to air, heat, cold, rain, and waves on the receding tide. Intertidal organisms have evolved various ways to deal with such stresses. For example, small marine snails, or periwinkles, that live in the rocky intertidal zone of the tropics, have various ways to avoid overheating when the tide is out. They have light-coloured shells to reduce heat absorption, little bumps on their shells that act like the cooling fins on a radiator, and they hang on to the rocks as much as possible with a mucous thread to avoid direct contact with the hot substratum. Mussels and barnacles living in the rocky intertidal avoid water loss and desiccation at low tide by tightly closing their shells and trapping enough water inside to survive until the next high tide. Crabs seek shelter in crevices or under moist mats of seaweed, or simply retreat down the shore with the receding tide. Some seaweeds living in the intertidal can tolerate extremes of dehydration, losing up to 90 per cent of the water in their tissues at low tide. Others secrete a gelatinous mucous covering that helps lock in water. Intertidal animals and plants living in cold climates must cope with extremes of cold temperature, sometimes tens of degrees below freezing, for periods of hours or days when exposed to the air. Some of these organisms produce antifreeze compounds that help prevent their tissues from freezing or are otherwise extremely tolerant to becoming frozen.

Intertidal organisms on exposed shorelines must also deal with the crushing and dragging forces of waves. Adult barnacles and oysters deal with these forces by cementing themselves permanently to rocks, whilst limpets, snails, and chitons hang on tightly with muscular foot-like attachment structures; adult mussels secure themselves to rocks by secreting strong, thread-like fibres called 'byssal threads'. Intertidal algae use holdfasts to attach to a hard surface and have flexible fronds that are able to bend and twist with the waves without damage.

Intertidal zonation

A classic feature of the intertidal region, particularly on rocky shores, is vertical zonation—the separation of intertidal life into prominent horizontal bands, often distinguished by different colours (see Figure 30).

The observed pattern of vertical zonation on rocky shores is quite similar from one region to another around the planet, which has led to a generally accepted 'universal' system for

30. **Typical pattern of intertidal zonation at low tide on a rocky shore in Washington State, USA**

describing these zones. In this system the intertidal region is divided into four zones which, from highest to lowest on the shore, can be referred to as the splash zone, the high intertidal zone, the mid intertidal zone, and the low intertidal zone (see Figure 31).

The splash zone is above the line of highest spring tides and its only direct connection with the marine environment is the spray from waves. This is a sparsely populated part of the shore because few organisms can withstand its extremely harsh conditions. The high intertidal zone is completely submerged only during spring tides, with parts exposed to the air for days to weeks, so it too is quite a harsh environment. The mid intertidal zone is the area of the shore between about the average high tide and the average low tide levels, and thus most of it is submerged for prolonged periods during most tide cycles. This zone is densely covered by a variety of marine plants and animals. The low intertidal zone is the area between the average low tide level and lowest spring tides. Thus it

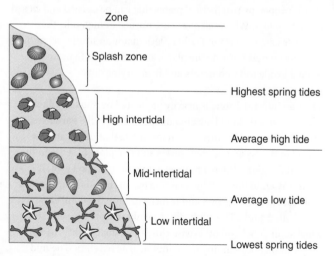

31. The division of the rocky intertidal into four 'universal' zones

stays completely submerged during most tide cycles and is the least physically stressed part of the intertidal region.

On rocky intertidal shores in temperate regions the splash zone is colonized by patches of bright orange- and greyish-coloured lichens; blue-green algae (cyanobacteria) that form a thin black layer on the rocks; and hair-like green algae. Periwinkles and limpets are common in the splash zone, grazing on the algae, along with isopods scavenging on dead organic material. For these animals, the splash zone is a refuge from predatory crabs and snails that cannot venture for long into this harsh environment.

The high intertidal zone is dominated by barnacles which are often so densely packed together that they create a distinct white band on the shore. Barnacles have a tiny, shrimp-like larval stage that drifts and swims about freely in the ocean. When the larva finds a suitable place to settle in the intertidal region, it secretes glue from a gland in its head and cements itself onto a solid surface. It then secretes a calcareous box-like shelter around itself that is capped by two pairs of plates that can be opened and closed like a trapdoor. When the plates are opened at high tide, the barnacle extends a set of feathery legs that filter plankton out of the seawater. At low tide the plates are sealed shut to protect the animal inside from predators and from drying out.

The mid intertidal zone is heavily populated by mussels which often create a distinct black band on the shore. Like barnacles, mussels also produce tiny free-living larvae that disperse for a period in the ocean and then seek a suitable place to settle in the intertidal region. When a mussel larva detects a good spot, it secures itself to the rock by secreting byssal threads from a gland in its foot and then grows and transforms into the familiar two-shelled adult. The adults filter plankton from the seawater when submerged and otherwise close their shells tightly to avoid dehydration and roaming predators such as sea stars. The mid intertidal zone is also occupied by oysters, limpets, and periwinkles as well as various species of fleshy brown seaweeds which provide a moist shelter for sea stars, sea urchins, and other marine animals at low tide.

The low intertidal zone is richly populated by a range of plants and animals that can tolerate occasional exposure to air. Here, red, green, and brown seaweeds proliferate, along with many different kinds of marine animals including sea anemones, sea stars, sea urchins, brittle stars, sea cucumbers, crabs, snails, sea slugs, and worms.

On rocky intertidal shores in the tropics, various species of algae create grey- and black-coloured films on the rocks in the splash zone. Here, a variety of periwinkles graze on these films. Rock-boring algae often give the high intertidal zone of the tropics a yellow appearance. Barnacles, limpets, and snails can be present in the high intertidal, but generally in small numbers compared to temperate rocky shores. A pink zone is often present in the mid intertidal zone created by encrusting coralline algae. Various snails occupy this zone along with mussels, sea anemones, limpets, and barnacles. Brown seaweeds often cover rocks in the low intertidal zone, which is home to a variety of marine organisms such as sea urchins, sea anemones, limpets, sea cucumbers, and sponges.

Causes of intertidal zonation

For many decades marine biologists have worked at understanding the factors that create the distinct patterns of vertical zonation on rocky intertidal shores. It is now known that intertidal zonation is caused by a complex interaction of biological and physical factors. In general, a plant's or animal's tolerance to physical factors, such as exposure to air, heat, cold, desiccation, and wave forces, determines where in the intertidal region it can potentially live. However, biological factors, such as competition, grazing, predation, and patterns of larval settlement, interact and modify the influence of physical factors, and ultimately determine where an organism actually occurs on the shore.

The biological interactions between barnacles, mussels, sea stars, and seaweeds on a temperate rocky shore provide a good example of how this works. Living space is at a premium in the intertidal region and hence competition for space is an important factor influencing the actual distribution of some species on the shore. Given a sufficient supply of larvae, barnacles have the ability to settle and live anywhere in the mid and high intertidal zones of temperate rocky shores because of their tolerance to physical stresses. However, they are generally excluded from much of the more benign mid intertidal zone where they would be subject to less stress, have more time to feed and grow faster. This is because mussels outcompete barnacles for space in the mid intertidal. They do this by overgrowing and smothering the barnacles. Thus barnacles ultimately persist mainly in the refuge of the high intertidal zone where mussels cannot tolerate the harsh environment.

However, barnacles are not entirely excluded from the mid intertidal and persist in patches here and there in this zone. This is a result of another biological interaction—predation. In the mid intertidal, mussels can be eaten by sea stars and are thus prevented from completely dominating this zone. However, sea stars can only venture up into the mid intertidal to eat mussels for limited periods at high tide, and thus are prevented from overeating mussels. In the low intertidal zone, though, sea stars can prey on mussels mostly at will and so almost completely eliminate mussels in this zone.

Hence, in this situation, the upper limit of barnacle distribution is set by physical factors and the lower limit by a biological factor—competition for space from mussels; similarly, the upper limit of mussel distribution is set by physical factors and the lower limit by a biological factor, in this case predation by sea stars.

The more closely the rocky intertidal region is studied, the greater the complexity of biological interactions revealed. For example, brown seaweeds are also competing for space in the mid intertidal and, if they become well established, they can maintain their foothold by preventing mussel and barnacle larvae from settling there. This appears to be a result of the blades of the seaweeds being swept back and forth across the rocky surface by wave action and preventing mussel and barnacle larvae from securing an attachment. On the other hand, grazing on seaweeds by limpets and snails can reduce seaweed cover, allowing mussels and barnacles to become established. In response to grazing, some seaweeds are able to produce toxic chemicals that deter further grazing.

Human impacts

The intertidal region can be subject to severe human impact on account of its accessibility and very close association with land and human development. It supplies humans with a ready source of wild marine food such as mussels, oysters, limpets, periwinkles, sea urchins, crabs, clams, abalone, and various seaweeds which are often harvested on a recreational and largely unregulated basis. In some places close to populated regions, such organisms have been overharvested and are now scarce or absent in the intertidal where they were once abundant. Such harvesting can also fundamentally change the structure of intertidal communities. A good example is provided by the harvesting of a large predatory intertidal snail by local people in Chile. In areas where this snail is harvested, the mid intertidal zone is dominated by a monoculture of mussels. However, where such harvesting is prevented, the mid intertidal is populated by barnacles and seaweeds and has a higher overall species diversity. This is because the snail eats mussels and so prevents the mussels from completely outcompeting other species for space. In some jurisdictions around the world, steps have been put in place to more carefully manage the impacts of recreational harvesting in the intertidal by enforcing recreational quotas or by establishing 'no take' marine reserves where harvesting of any kind is not allowed.

Larger-scale commercial harvesting of intertidal organisms also occurs. For instance, an abundant brown seaweed, known as 'rockweed', is harvested on a commercial basis from intertidal regions of eastern Canada and Maine. The seaweed is dried and used for fertilizer, as an animal feed, and for the extraction of alginates which have many uses, including as an additive to ice creams and other dairy products. Wild mussels are also harvested commercially off the coast of Maine, while oysters are harvested commercially in the intertidal regions of South Africa and elsewhere around the world.

Just the simple act of walking and turning over rocks in the intertidal to observe the organisms underneath can be destructive. This is not an issue on isolated coasts, but rocky shores near urban centres attract huge numbers of visitors and can be highly impacted. To put this into perspective, some popular 'high use' rocky intertidal sites along the Californian coast attract 25,000 to 50,000 visitors per year per 100 metres of shoreline. Less popular sites are still subjected to 2,000 to 10,000 visitors per year per 100 metres of shoreline. The trampling caused by all these visitors walking in the intertidal at low tide dislodges and crushes plants and animals. Many visitors also turn over rocks, which will crush the organisms living on the top of the rock and expose those organisms living under the rocks to desiccation, wave action, and predation. This creates 'monk's head rocks' which have only a fringe of organisms around their edges and none on the top and bottom surfaces. Even in no-take marine reserves, people can walk freely in the intertidal and so this kind of impact will still occur. Thus, if there is a desire to more fully protect some high-use intertidal regions, it may be necessary to limit the number of visitors or keep them to intertidal 'trails'.

In our fossil-fuelled civilization, oil pollution is a threat to intertidal communities just about anywhere on the planet, as evidenced by the ubiquitous tar balls—balls of crude oil—often found washed up on beaches. Crude oil gets into the ocean from four major sources: natural seeps, extraction processes, transportation, and consumption.

Seepage of crude oil into the marine environment occurs naturally from oil-containing geological formations below the seabed. It is estimated that around 600,000 tonnes of crude oil seeps into the marine environment each year, which represents almost half of all the crude oil entering the oceans. This large amount of crude oil is, however, released at a slow enough rate, and from so many different locations, that the marine environment is not damaged.

The human activities associated with exploring for and producing oil result in the release on average of an estimated 38,000 tonnes of crude oil into the oceans each year, which is about 6 per cent of the total anthropogenic input of oil into the oceans worldwide Although small in comparison to natural seepage, crude oil pollution from this source can cause serious damage to coastal ecosystems because it is released near the coast and sometimes in very large, concentrated amounts. In fact, the catastrophic Gulf of Mexico oil spill of 2010 is so far the largest accidental release of oil into the marine environment in history. In this incident, a deep-ocean drilling platform, which was drilling an exploratory well about 66 kilometres off the coast in about 1,500 metres of water, failed, allowing huge amounts of crude oil to gush from a reservoir under the seabed (see Figure 32). During the course of a three-month period, about 670,000 tonnes of oil escaped into the Gulf of Mexico before the undersea gusher was finally capped. Around 790 kilometres of coastline were contaminated by oil which coated, smothered, and poisoned intertidal and subtidal marine life as well as coastal wildlife. It also resulted in the closure of shrimp fisheries in much of the Gulf. In addition, the chemical

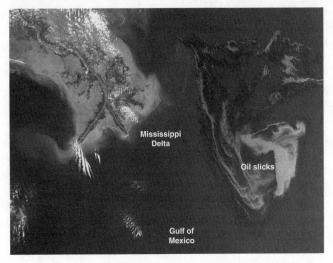

32. Satellite image of oil slicks from the Gulf of Mexico oil spill

dispersants and mechanical devices used in the clean-up phase caused further damage to marine life.

The transport of oil and oil products around the globe in tankers results in the release of about 150,000 tonnes of oil worldwide each year on average, or about 22 per cent of the total anthropogenic input. Oil spills from tankers can be catastrophically large and concentrated, and thus very damaging. When the tanker *Exxon Valdez* grounded on rocks off the coast of Alaska in 1989, at least 30,000 tonnes of oil were spilled into Prince William Sound. Currents carried the oil slick down the Alaskan coast, where it coated around 2,100 kilometres of coastline and extensively damaged the intertidal and killed hundreds of thousands of seabirds, thousands of marine mammals, and an indeterminate number of fish. The impacts of this natural disaster linger to this day and full recovery of the contaminated areas could take as much as thirty years.

About 480,000 tonnes of oil make their way into the marine environment each year worldwide from leakage associated with the consumption of oil-derived products in cars and trucks, and to a lesser extent in boats. Oil lost from the operation of cars and trucks collects on paved urban areas from where it is washed off into streams and rivers, and from there into the oceans. Surprisingly, this represents the most significant source of human-derived oil pollution into the marine environment—about 72 per cent of the total. Because it is a very diffuse source of pollution, it is the most difficult to control. The impact of this chronic, diffuse oil pollution on marine organisms and the functioning of marine communities is insidious. It is known that the many different kinds of organic hydrocarbons found in oil have deleterious and cumulative effects on many marine organisms, even at very low concentrations, with the larval stages of marine animals being particularly vulnerable.

Chapter 8
Food from the oceans

Humans have been harvesting food from the oceans for millennia. It is often assumed that in ancient times catches of fish and other marine seafood were limited by low population numbers and the simplicity of fishing gear and vessels available, and that seafood stocks were only lightly exploited until recent times. However, it has been deduced from historical marine ecological studies that humans have had significant impacts on marine resources for thousands of years. The Mediterranean Sea provides a good example of long-standing human pressure on marine resources.

Humans have occupied the shores of the Mediterranean Sea on a sustained basis for about 50,000 years. Fishing communities have existed there for at least 10,000 years and seafood has been an important source of protein for Mediterranean people since at least the Greek and Roman periods, starting about 900 BC. Species fished included dolphins, sea turtles, sharks, rays, tuna, sardines, anchovies, mullet, grouper, flatfish, oysters, mussels, clams, and scallops. Some of these species were depleted by Roman times and by the 1st century AD the coastal waters around Italy had been largely overfished and harvesting had spread to offshore islands like Sicily and Corsica.

In the late 19th century, fishing capacity grew greatly in the Mediterranean as population numbers in the region rapidly expanded. Through the early and mid 20th century, the fishing fleet became motorized and fishing expanded into offshore waters of the Mediterranean Sea. Today it has been estimated that virtually all of the marine food resources in the Mediterranean Sea have been reduced to less than 50 per cent of their original abundance and about a third are now very rare. The greatest impact has been on the larger predatory fish, which were the first to be targeted by fishers.

The intensive exploitation of marine resources got under way in other parts of Europe much later than in the Mediterranean. However, by about the 8th or 9th centuries, the Vikings were exploiting large stocks of cod, haddock, pollock, herring, and other marine species that flourished in the seas of northern Europe at that time. They brought this skill with them when they spread into Britain and Normandy, where marine fishing became well established by the 11th century. By the end of the 18th century, many of the fish stocks of northern Europe were in serious decline from overfishing and the first steps were taken to protect some stocks. Nonetheless, intensive fishing continued more or less unabated and by 2000 more than half of northern European fish stocks were considered seriously overexploited. It is estimated that, collectively, the European fish stocks of today are just one-tenth of their size in 1900. Going back to a time when European fish stocks were unexploited, it has been suggested that at present European seas contain less than 5 per cent of the total mass of fish that once swam there.

Exploitation of marine food resources spread outwards from Europe surprisingly early in history. As early as 1000 AD, Vikings were exploiting the rich fish stocks around Iceland and Greenland and possibly into the seas of northern Canada. Some 200 years later, Basque fishers were also harvesting North Atlantic fish stocks and may have been fishing off the coast of North America before Columbus's voyage of 1492. Nonetheless, by the early 1500s, Basque, French, and Portuguese fishers were routinely exploiting cod from the shallow Grand Banks off the coast of Newfoundland for the European market. By 1600, more than 150 shiploads of cod per year were being extracted from Canadian seas. The catch was preserved by either drying or salting it and in this form it could be stored for several years and provided a ready source of protein for Europeans. The permanent colonization of North America starting in the 17th century marked the beginning of the exploitation of the virgin fish stocks off the east coast of the United States, including sturgeon, shad, salmon, alewife, and oysters. By the early 1800s, the fish stocks of the eastern seaboard of North America were showing early signs of serious decline.

In the late 1800s and early 1900s, rapidly increasing urbanization and global population growth created an ever-increasing demand for seafood. This was linked to improvements in transport and preservation of the catch that meant that offshore fish stocks could be more easily exploited and distant markets served. Globally, the marine harvest increased at an unprecedented rate in size and geographic footprint after the Second World War, with a rapid increase in the size of the global fishing fleet, and improvements in fishing gear and the introduction of onboard processing of the catch. This marked the start of the mass industrialization of extraction of food from the sea that penetrated just about every marine habitat on the planet.

The advent of motorized fishing vessels enabled super-efficient fish catching with the use of trawl nets. Trawls are large, cone-shaped nets that are towed along the seabed, scooping up fish and other marine organisms in their path (see Figure 33). Large trawl doors made of metal or wood are attached to the net with wire bridles and hold the net open and help to herd fish into the net. The mouth of a large trawl net can be about the size of a football field. Steel or rubber discs, called bobbins, are attached to a foot rope on the lower lip of the trawl net and allow the trawl to ride over rough bottom. Commercial trawlers ply the shallow

33. Fishers unloading a trawl net full of orange roughy

waters of the continental shelves and can also now fish the continental slopes as well as deep-sea canyons and seamounts down to depths of over 2,000 metres.

On heavily fished areas of the continental shelves, the same parts of the sea floor can be repeatedly trawled many times per year. Such intensive bottom trawling causes great cumulative damage to seabed habitats. The trawls scrape and pulverize rich and complex bottom habitats built up over centuries by living organisms such as tube worms, cold-water corals, and oysters. These habitats are eventually reduced to uniform stretches of rubble and sand. For all intents and purposes these areas are permanently altered and become occupied by a much changed and much less rich community adapted to frequent disturbance.

There are thousands of species of marine fish but the majority of the global catch today comprises a small number of species falling

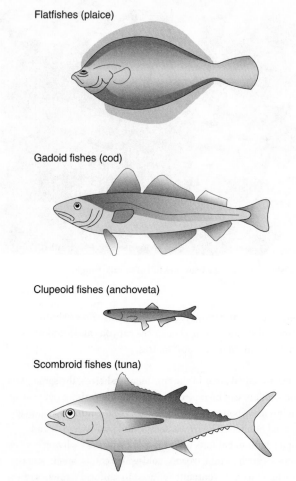

Flatfishes (plaice)

Gadoid fishes (cod)

Clupeoid fishes (anchoveta)

Scombroid fishes (tuna)

34. Major groups of commercial marine species

into several main groups (see Figure 34). The group that accounts
for the greatest proportion of the marine catch is the clupeoid fish.
These are small, schooling fish that feed directly on phytoplankton
and zooplankton and include the herrings, sardines, and

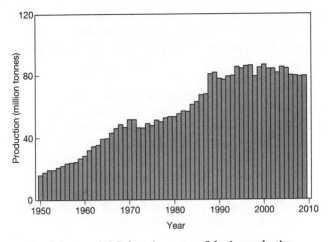

35. Trends in annual global marine capture fisheries production

anchovies. The gadoid fish are another important group of
commercial species. These are bottom-dwelling fish and include
cod, haddock, hake, and pollock, which are species that live in
coastal waters of the North Pacific and North Atlantic Oceans.
Flatfish such as flounder, halibut, sole, and plaice are another
important group of commercial species, whose habitat is also in
coastal waters. Then there are the large, fast-swimming, pelagic
carnivores—the scombroids, which include mackerels and tunas,
the latter a major open-ocean fishery. Sharks comprise another
exploited group of marine fish. There are also large fisheries
for marine invertebrates, including crustaceans such as
lobsters, crabs, and shrimp; and molluscs such as squid, oysters,
and clams.

In 1950 the total global catch of marine seafood was just less
than twenty million metric tonnes fresh weight. This increased
steadily and rapidly until by the late 1980s more than eighty
million tonnes were being taken each year (see Figure 35).
Starting in the early 1990s, however, yields began to show signs

of levelling off. The highest catch on record was about eighty-six million tonnes in 1996. Since then catches have fluctuated at below this level and the catch in 2009 was about 79.9 million tonnes.

By far the most heavily exploited marine fishery in the world is the Peruvian anchoveta (*Engraulis ringens*) fishery, which can account for 10 per cent or more of the global marine catch of seafood in any particular year. Anchoveta are small (up to about 20 cm in length) fast-growing fish that filter-feed on phytoplankton and zooplankton in the nutrient-rich upwelling areas off the coast of Peru. They tend to congregrate in large, dense schools, which means they can be caught efficiently in large numbers by purse seiners and trawlers.

The anchoveta catch fluctuates wildly from year to year. In a good year catches can be upwards of eleven million tonnes; in bad years the catch can be around two million tonnes; and in really bad years below 150,000 tonnes. Much of this fluctuation is related to the El Niño Southern Oscillation (ENSO), whose impact is exacerbated by overfishing. During El Niño events the Pacific trade winds blowing out of the east diminish, allowing warm tropical surface water to flow eastwards across the Pacific and accumulate along the coast of South America. This pool of warm surface seawater effectively shuts down the upwelling of nutrient-rich seawater along the coast of Peru, leading to a drastic decrease in phytoplankton populations, the main food of the anchoveta, which then die off. In turn, seabirds and predatory fish that depend on the anchoveta for food suffer massive mortalities. Severe El Niño events occurred in 1972–3, 1976, 1982–3 and 1997–8, and these periods were all associated with dramatic decreases in the amount of available fish and the resultant catches. Unfortunately, overfishing during these El Niño periods compounded the natural impact on the anchoveta population and resulted in declines in future catches.

The anchoveta is a very oily fish, which makes it less desirable for direct consumption by humans. However, the high oil content makes it ideal for the production of fish meal and fish oil, which are both very valuable commodities. Fish meal is made by cooking, drying, and grinding up the fish and is used as a protein supplement incorporated into commercial feeds for poultry, pigs, and farmed fish. Fish oil is pressed from the cooked fish and is used mainly in the production of feed for farmed fish, but is also put into capsules as a human health supplement. Almost the entire Peruvian anchoveta catch is converted into fish meal and fish oil, which accounts for about a third of the global production of these products. However, the demand for fish meal and oil is huge and about a third of the entire global catch of fish is converted into these products rather than consumed directly by humans. Feeding so much fish protein to livestock comes with a considerable loss of potential food energy (around 25 per cent) compared to if it was eaten directly by humans. This could be viewed as a potential waste of available energy for a rapidly growing human population that is quickly reaching the limits of food production. However, around 90 per cent of the fish used to produce fish meal and oil is presently unpalatable to most people and thus unmarketable in large quantities as a human food, which explains why so much fish is turned into livestock feed at present.

The eighty million tonnes or so of marine seafood caught each year globally equates to about eleven kilograms of wild-caught marine seafood per person on the planet. Wild-caught seafood is a common part of the diet of many billions of people worldwide and seafood provides more than 1.5 billion people with at least 15 per cent of their average per capita protein requirements. Looking ahead, what contribution will the Global Ocean make to providing food for the nine billion people that will occupy our planet by 2050?

One way to approach this question is founded on basic ecological principles around the transfer of energy between trophic levels in the oceans, a trophic level being a step in the food chain. John Ryther was the first to do this back in 1969 when he estimated the total fish production of the Global Ocean based on its total primary production (see Table 2). Ryther divided the ocean into three provinces designated by their different levels of primary productivity. The largest, the oceanic province, consists of all of the open-ocean regions of the Global Ocean. It has the lowest average primary productivity but occupies the greatest area by far—about 90 per cent of the total area of the Global Ocean. The second is the coastal province encompassing the coastal marine waters of the Global Ocean. It has higher primary productivity but a much smaller area—just under 10 per cent. The third province he designated as the upwelling province and includes all of the major upwelling regions of the Global Ocean. It occupies a tiny area—about 1 per cent of the Global Ocean—but has very high average primary productivity.

Ryther then estimated the number of trophic levels in each province between the primary producers and the major fish species captured in each province. Because food webs in the open ocean are complex, he reasoned that there were on average five trophic levels between the primary producers and the exploited fish. By comparison, food webs in upwelling regions are very simple. The Peruvian anchoveta fishery is a case in point—here the anchoveta feed directly on phytoplankton or on zooplankton, so they are only one or two trophic levels removed from the primary producers. Thus Ryther assigned this province 1.5 trophic levels between primary producers and fish. Ryther reasoned that the coastal province was transitional in nature between the open ocean and the upwelling regions and assigned it three trophic levels.

Ryther then went back to first principles and estimated the efficiency with which energy is transferred between each trophic level in each province. He knew that energy transfer would be least efficient in the complex food webs of the open ocean and assigned it a value of 10 per cent. However, energy transfer would

Table 2 Estimated total fish production of the Global Ocean

Province	% of the ocean	Mean productivity ($g C m^{-2} yr^{-1}$)	Total production (billions tonnes of $C yr^{-1}$)	Number of trophic levels	% efficiency	Fish production (tonnes fresh weight)
Oceanic	90.0	50	16.3	5	10	1.6 million
Coastal	9.9	100	3.6	3	15	120 million
Upwelling	0.1	300	0.1	1.5	20	120 million
Total						Approximately 240 million

be much more efficient in the simple food webs of the upwelling zones, and here he assigned a figure of 20 per cent. He assigned the intermediate figure of 15 per cent efficiency of energy transfer for the coastal province.

Ryther then used the information that was available to him at the time on primary productivity in each province to estimate the total yearly primary production in each province. By combining all of the data on primary production, trophic levels, and energy transfer efficiencies, he came up with a figure for total world commercial fish production of 240 million metric tonnes per year.

One of the insights from Ryther's analysis is the overwhelming importance of coastal and upwelling regions in commercial fish production because of their high primary productivity and simpler food webs. Together they account for almost all of the exploitable fish production of the Global Ocean. The open-ocean region, despite its immense size, accounts for little commercial fish production. This is consistent with the known facts—the largest commercial fisheries—for clupeoids, gadoids, and flatfish—are all situated in upwelling regions and shallow coastal parts of the Global Ocean. The exception is the open-ocean fisheries for species such as tuna and mackerel, the top predators in a vast oceanic system.

Not all of the estimated 240 million tonnes of fish production is available to humans for harvest. If all of it were exploited there would be no breeding stock for future harvests. Also, much of this fish production is consumed by other top predators besides humans—such as marine mammals and seabirds. On this basis, Ryther estimated that a maximum of about a hundred million tonnes of fish would be available for exploitation by humans on a sustained basis each year.

In 1971, J. A. Gulland, using a similar approach to Ryther's, also estimated the potential maximum fish harvest from the Global

Ocean, not including invertebrates, as close to a hundred million tonnes per year. However, he reasoned that it would be impossible to exploit all fish stocks at an optimal level at all times and suggested that a figure of eighty million tonnes was more realistic.

These important studies have provided a theoretical underpinning for estimating the amount of fish that humans might expect to be able to sustainably harvest from the Global Ocean based on ecological principles. In reality, the global marine fish catch has flattened out at under eighty million tonnes per year, suggesting that the theoretical approaches have overestimated the actual amount of fish that humans might expect to harvest from the Global Ocean under ideal circumstances. What is perfectly clear, though, on the basis of theory backed up by real data on marine fish catches, is that marine fisheries are now fully exploited and that there is little if any headroom for increasing the amount of wild-caught fish humans can extract from the oceans to feed a burgeoning human population. The 70 per cent or so extra food needed to feed nine billion mouths in the near future will have to come from somewhere else.

This conclusion is solidly supported by the increasingly precarious state of global marine fishery resources. The most recent information from the Food and Agriculture Organization of the United Nations (*The State of World Fisheries and Aquaculture 2010*) shows that over half (53 per cent) of all fish stocks are fully exploited—their current catches are at or close to their maximum sustainable levels of production and there is no scope for further expansion. Another 32 per cent are overexploited and in decline. Of the remaining 15 per cent of stocks, 12 per cent are considered moderately exploited and only 3 per cent underexploited. Over the past forty years there has been a drastic decrease in moderately exploited or unexploited fish stocks—in the mid 1970s 40 per cent of all fish stocks were in this category as opposed to around 15 per cent now.

All in all this paints a grim picture of the state of our planet's marine fisheries. The majority of stocks are overfished—and thus face imminent collapse—or are fully exploited and cannot be expanded. And there are very few moderately exploited or underexploited fish stocks left. Thus the real question is not so much whether we can get more fish from the sea but whether we can sustain the amount of fish we are harvesting at present, which represents a vital source of food for the human family.

There is no firm consensus on this question yet but there is cause for pessimism. It will be very difficult to control overexploitation and serial collapse of fish stocks in the face of the ever-increasing demand for seafood fuelled by a rapidly growing global human population. Governments around the world will have to work harder and more assiduously towards measures that will reduce fishing effort to levels that will allow long-term sustainability of fish stocks, and also be prepared to make the resources available to enforce these measures. They will also have to make hard and unpopular decisions to completely close fisheries that are not managed in a sustainable fashion. The retailers and consumers of seafood have an important role to play here as well. Overexploitation of fish stocks is often driven by consumer demand and if consumers are properly educated and knowledgeable about the problems of overfishing they will be in a position to make decisions not to purchase overexploited species and to support sustainable fishery products. Charts are now available in some stores that list fish obtained from properly managed fisheries and those from overfished and unmanaged fisheries. The consumer is then in a position to make better choices.

A vital step towards restoring marine fisheries is the creation of a network of marine reserves, an initiative eloquently argued by Callum Roberts in his book *The Unnatural History of the Sea*. We take national and regional parks and protected areas on land for granted as a rational requirement for preservation of significant and representative areas of terrestrial wildlife and landscape in the face of growing human pressure. At this time about 12 per cent of the planet's land area is now under some form of protection. The corresponding figure for the oceans is well less than 1 per cent, with most of this area still open to some form of fishing. The area of the oceans where human exploitation is completely restricted is miniscule and consists of a small number of scattered no-take marine reserves.

Marine reserves can be used to preserve representative habitats in the oceans that are still relatively intact. They can also be used to allow impacted environments to begin to recover so as to recreate more normally functioning marine systems. They are also beneficial to the recovery of fisheries. Marine reserves afford a protected area within which commercial species of fish and invertebrates have a chance to avoid being captured and thus live longer and reach a larger body size. Larger body size equates to a greater number of eggs being produced, while the longer an animal lives, the more times it can reproduce. In this way, local fisheries can begin to recover. There is also a 'spillover' effect created by marine reserves. As populations of various mobile species such as fish and lobsters begin to recover and proliferate within a marine reserve, they eventually reach a point where they saturate the protected area and begin to spread out into unprotected areas where they bolster recreational and commercial fisheries. Furthermore, the dense populations of large-bodied animals within the reserve are able to produce large numbers of eggs and larvae that drift out of the reserve and enhance recruitment in adjacent areas. It is not surprising, then, as many fishers have come to realize, that the best fishing is often adjacent to an established marine reserve.

A key question becomes how much of the Global Ocean we would need to protect in order to allow sufficient overexploited marine systems to recover to achieve a more sustainable marine harvest. The consensus among marine scientists working in this area is that somewhere in the vicinity of 20 to 40 per cent of the oceans need to be protected to maximize return to the fishing industry. This means we would need roughly fifty times the area presently under protection scattered widely across coastal seas and the open ocean. Some nations are moving boldly to embrace this vision. For example, South Africa already protects about 18 per cent of its coastal waters and is expanding its marine reserve system offshore; and in Australia 100,000 square kilometres of the Great Barrier Reef are protected from fishing. However, much more needs to be done at a much faster pace. The estimated costs of establishing and administering a global network of marine reserves are high but more than offset by the expected uplift in yield of commercial fisheries, the large numbers of jobs created, and the increased income from visitors.

Marine reserves will not address all the problems of the oceans. Apart from overfishing, marine ecosystems are under pressure from chemical pollution, agricultural run-off, sedimentation, and climate change effects such as ocean warming and acidification, all of which ignore the boundaries of marine reserves and work synergistically to reduce the ability of marine ecosystems to function normally. To address these issues will require a globally holistic approach to managing the environment of our planet and a fundamental reassessment of how we as humans interact with and value the natural systems of our planet and how we use its resources. This is human society's greatest challenge over the next twenty to thirty years as we climb steadily towards a population of nine billion, and the decisions made will determine the future of the Global Ocean, the largest and most important habitat on the planet.

Further reading

Useful textbooks and overview articles

A. Berta, J. L. Sumich, and K. M. Kovacs, *Marine Mammals Evolutionary Biology*, 2nd edn (Burlington, MA: Academic Press, 2006)

Tom Garrison, *Oceanography: An Invitation to Marine Science*, 8th edn (Belmont, CA: Brooks, Cole, Thomson Learning, 2013)

P. J. Hogarth, *The Biology of Mangroves and Seagrasses* (New York: Oxford University Press, 2007)

George Karleskint, Jr., Richard Turner, and James W. Small, Jr., *Introduction to Marine Biology*, 4th edn (Belmont, CA: Brooks/ Cole, Cengage Learning, 2010)

James W. Nybakken and Mark D. Bertness, *Marine Biology: An Ecological Approach*, 6th edn (San Francisco, CA: Pearson Benjamin Cummings, 2005)

Callum Roberts, *The Unnatural History of the Sea* (Washington, DC: Island Press/Shearwater Books, 2007)

E. E. Ruppert, R. S. Fox, and R. B. Barnes, *Invertebrate Zoology, A Functional Evolutionary Approach*, 7th edn (Belmont, CA: Cengage Learning, 2004)

Articles on more specialist topics in marine biology

Anonymous, *Land use and the Great Barrier Reef: current state of knowledge*, revised edn (Townsville, Queensland: QLD CRC Reef Research Centre, 2003)

Anonymous, *Oil in the Sea III: Inputs, Fates, and Effects* (National Academies Press, 2003) (http://www.nap.edu/catalog/10388.html)

Anonymous, *The State of World Fisheries and Aquaculture 2010* (FAO Fisheries and Aquaculture Department, Food and Agriculture Organization of the United Nations, Rome (2010))

Michelle Allsopp, Adam Walters, David Santillo, and Paul Johnston, *Plastic Debris in the World's Oceans* (44 pages) (Greenpeace International, 2006)

Donald M. Anderson, Patricia Glibert, and JoAnn M. Burkholder, *Harmful algal blooms and eutrophication: nutrient sources, composition and consequences*, Estuaries, 25 (2002), No 4b, pp. 704–26

A. Atkinson, V. Siegel, E. Pakhomov, and P. Rothery, *Long-term decline in krill stock and increase in salps within the Southern Ocean*, Nature, 432 (2004), pp. 100–3

Mya Breitbart, Luke R. Thompson, Curtis A. Suttle, and Matthew B. Sullivan, *Exploring the vast diversity of marine viruses*, Oceanography, 20 (2) (2007), pp. 135–9

J. C. Castilla, and L. R. Duran, *Human exclusion from the rocky intertidal zone of central Chile: the effects on Concholepas concholepas (Gastropoda)*, Oikos, 45 (1985), pp. 391–9

M. R. Clark, D. Tittensor, A. D. Rogers, P. Brewin, T. Schlacher, A. Rowden, K. Stocks, and M. Consalvey, *Seamounts, deep-sea corals and fisheries: vulnerability of deep-sea corals to fishing on seamounts beyond areas of national jurisdiction.* (UNEP-WCMC, Cambridge, UK (2006))

Maj De Poorter, C. Darby, and J. MacKay, *Marine menace: alien invasive species in the marine environment* (Gland, Switzerland: IUCN, 2009), pp. 1–31

Robert J. Diaz and Rutger Rosenberg, *Spreading dead zones and consequences for marine ecosystems*, Science, 321 (5891) (2008), pp. 926–9

N. C. Duke, J.-O. Meynecke, S. Dittmann, A. M. Ellison, Klaus Anger, U. Berger, S. Cannicci, K. Diele, K. C. Ewel, C. D. Field, N. Koedam, S. Y. Lee, C. Marchand, I. Nordhaus, and F. Dahdouh-Guebas, *A World Without Mangroves?*, Science, 317 (5834) (2007), pp. 41–2

James J. Elser et al., *Global analysis of nitrogen and phosphorus limitation of primary producers in freshwater, marine and terrestrial ecosystems*, Ecology Letters, 10 (2007), pp. 1–8

Paul G. Falkowski, *The Ocean's Invisible Forest*, Scientific American, 287, August 2002, pp. 54–61

Gabriel M. Filippelli, *The Global Phosphorus Cycle: Past, Present, and Future*, Element, 4 (2008), pp. 89–95

J. A. Fuhrman, *Marine viruses and their biogeochemical and ecological effects*, Nature, 399 (1999), pp. 541–8

J. A. Gulland, ed., *The fish resources of the ocean*, (West Byfleet, UK, Fishing News (Books) Ltd (1971))

J. B. C. Jackson, *Ecological extinction and evolution in the brave new ocean*, Proceedings of the National Academy of Science of the USA, 105 Suppl. 1 (2008): pp. 11458–65

J. B. C. Jackson, *The future of the oceans past*, Philosophical Transactions of the Royal Society of London, B 365 (2010): pp. 3765–8

J. B. C. Jackson, *What was natural in the coastal oceans?*, Proceedings of the National Academy of Science of the USA, 98 (10) (2001): pp. 5411–8

Jeremy B. C. Jackson, Michael X. Kirby, Wolfgang H. Berger, Karen A. Bjorndal, Louis W. Botsford, Bruce J. Bourque, Roger H. Bradbury, Richard Cooke, Jon Erlandson, James A. Estes, Terence P. Hughes, Susan Kidwell, Carina B. Lange, Hunter S. Lenihan, John M. Pandolfi, Charles H. Peterson, Robert S. Steneck, Mia J. Tegner, and Robert R. Warner, *Historical Overfishing and the Recent Collapse of Coastal Ecosystems*, Science, 293 (5530) (2001), pp. 629–37

Karl-Hermann Kock, *Antarctic marine living resources—exploitation and its management in the Southern Ocean*, Antarctic Science, 19 (2) (2007), pp. 231–8

W. Lampert, *The adaptive significance of diel vertical migration of zooplankton*, Functional Ecology, 3 (1989), pp. 21–7

Heike K. Lotze, Marta Coll, and Jennifer A. Dunne, *Historical changes in marine resources, food-web structure and ecosystem functioning in the Adriatic Sea, Mediterranean,* Ecosystems, 14 (2011), pp. 198–222

Loren McClenachan, Jeremy B. C. Jackson, and Marah J. H. Newman, *Conservation implications of historic sea turtle nesting beach loss*, Frontiers in Ecology and the Environment, 4 (2006), pp. 290–6

D. G. M. Miller, *Exploitation of Antarctic Marine Living Resources: A Brief History and a Possible Approach to Managing the Krill Fishery*, South African Journal of Marine Science, 10 (1991), pp. 321–39

Marjorie Mulhall, *Saving the rainforests of the sea: an analysis of international efforts to conserve coral reefs*, Duke Environmental Law and Policy Forum, 19 (2009), pp. 321–51

Further reading

E. Ramirez-Llodra, A. Brandt, R. Danovaro, E. Escobar, C. German, L. Levin, P. Martinez Arbizu, L. Menot, P. Buhl-Mortensen, B. Narayanaswamy, C. Smith, D. Tittensor, P. Tyler, A. Vanreusel, and M. Vecchione, *Deep, diverse and definitely different: unique attributes of the world's largest ecosystem*, Biogeosciences Discussions, 7 (2010), pp. 2361–485

A. D. Rogers and D. d'A. Laffoley, *International earth system expert workshop on ocean stresses and impacts, Summary report* (18 pages) (IPSO Oxford, 2011)

Stuart A., Sandin, Jennifer E. Smith, Edward E. DeMartini, Elizabeth A. Dinsdale, Simon D. Donner, Alan M. Friedlander, Talina Konotchick, Machel Malay, James E. Maragos, David Obura, Olga Pantos, Gustav Paulay, Richie Morgan, Rohwer Forest, Robert F. Schroeder, Sheila Walsh, Jeremy B. C. Jackson, Nancy Knowlton, and Sala Enric, *Baselines and degradation of coral reefs in the northern Line Islands*, PLoS ONE, 3 (2008), pp. 1–11

C. R. Smith and A. R. Baco, *Ecology of whale falls at the deep-sea floor*, in R. N. Gibson and R. J. A. Atkinson, eds, Oceanography and Marine Biology: An Annual Review, 41 (2003), pp. 311–54

J. Smith, P. Fong, and R. Ambrose, *The impacts of human visitation on mussel bed communities along the California coast: are regulatory marine reserves effective in protecting these communities?*, Environmental Management, 41 (4) (2008), pp. 599–612

Curtis A. Suttle, *Marine viruses—major players in the global system*, Nature Reviews Microbiology, 5 (2007), pp. 801–12

M. J. Tegner and P. K. Dayton, *Ecosystem effects of fishing in kelp forest communities*, ICES Journal of Marine Science, 57 (2000), pp. 579–89

A. R. Townsend and R. W. Howarth, *Fixing the Global Nitrogen Problem*, Scientific American, 302 (February 2010), pp. 50–7

Websites

http://www.mbl.edu
http://www.whoi.edu/oceanus/index.do
http://ioc-unesco.org/hab/
http://www.antarctica.gov.au/
http://www.gbrmpa.gov.au
http://www.mbari.org/default.htm
http://www.fao.org/fi/default_all.asp

Index

W

walrus 63
Wandering Albatross (*Diomedea exulans*) 72
Weddell seal 74
whale falls 108–10
whales 63, 69–74
whaling 73–4

wind-driven ocean circulation 14–15, 18
wood falls 109–10

Z

zooplankton 25, 33, 36, 111
zooxanthellae 81–2, 85, 94
Zostera 46

Expand your collection of
VERY SHORT INTRODUCTIONS